Equipment Management in the Post-Maintenance Era

A New Alternative to Total Productive Maintenance (TPM)

Equipment Management in the Post-Maintenance Era

A New Alternative to Total Productive Maintenance (TPM)

KERN PENG

CRC Press
Taylor & Francis Group
Boca Raton London New York

CRC Press is an imprint of the
Taylor & Francis Group, an **informa** business

A PRODUCTIVITY PRESS BOOK

CRC Press
Taylor & Francis Group
6000 Broken Sound Parkway NW, Suite 300
Boca Raton, FL 33487-2742

© 2012 by Taylor & Francis Group, LLC
CRC Press is an imprint of Taylor & Francis Group, an Informa business

No claim to original U.S. Government works

Printed in the United States of America on acid-free paper
Version Date: 20111122

International Standard Book Number: 978-1-4665-0194-2 (Hardback)

Library of Congress Cataloging-in-Publication Data

Peng, Kern.
    Equipment management in the post-maintenance era : a new alternative to total productive maintenance (TPM) / Kern Peng.
       p. cm.
    Includes bibliographical references and index.
    ISBN 978-1-4665-0194-2
    1. Industrial equipment--Maintenance and repair--Data processing. 2. Plant maintenance--Data processing. I. Title.

TS192.P46 2012
658.2'020285--dc23   2011045767

Visit the Taylor & Francis Web site at
http://www.taylorandfrancis.com

and the CRC Press Web site at
http://www.crcpress.com

This book is dedicated to my wife, Patricia, who is the coach of my pursuits, and to my two daughters, Ilena and Elena, who are the cheerleaders along my journey.

Contents

List of Figures xiii
List of Tables xv
Preface xvii
Acknowledgments xxi
About the Author.. xxiii

SECTION I   Introduction to
Equipment Management

Chapter 1   Introduction 3

    Background 3
    Maintenance Management 6
    Equipment Management...7
    Key Equipment Terminology..8
        Equipment States 8
        Equipment Time 8
        Equipment Actions..11
        Equipment Failure Patterns 12
        Equipment Performance Measurements............................14

Chapter 2   History of Equipment Management 17

    Introduction 17
    Phase 1: Breakdown Management 17
    Phase 2: Preventive Maintenance...18
    Phase 3: Productive Maintenance 20
    Phase 4: Total Productive Maintenance 22
    Phase 5: TPM with Predictive Maintenance..........................25
    Summary of the Pre-maintenance and Maintenance
    Phases...26

Chapter 3 Introduction to the Post-Maintenance Era 29

The New Business Environment 29
    Operational Changes... 30
    Equipment Characteristics..31
    New Enabling Technologies..33
    New Management Concepts 35
The Issues of Maintenance 36
    Problems in Objectives 37
    Structural Inefficiency...38
    Unsuitable for Changing Environment.............................39
Introduction to Phase 6: Post-Maintenance Era 40

SECTION II   The Maintenance System

Chapter 4 General Maintenance Concepts and Practices 45

Introduction 45
Preventive Maintenance 45
    Preventive Maintenance Schedule.................................47
    Preventive Maintenance Task List................................50
    Preventive Maintenance Staffing Requirements 52
    Preventive Maintenance Equipment/Tool
    Requirements 53
    Preventive Maintenance Materials/Parts Requirements... 54
    Preventive Maintenance Information System 54
Reliability-Centered Maintenance....................................55
Predictive Maintenance..58
Maintenance Prevention 61
Total Productive Maintenance 63
Terotechnology 65

Chapter 5 Maintenance Management Logistics 67

Introduction 67
Planning and Budgeting...67
    Strategic Planning in Maintenance............................... 68
    Headcount Plan..70
    Budget Plan..75
    Tactical Planning in Maintenance 82

Training and People Development 85
Customer Services and Management 92
Vendor, Supplier, and Contract Management 95
Inventory Management 100

Chapter 6 Maintenance Performance Indicators 105

Introduction 105
Equipment Performance Indicators..................................106
    Safety Indicators 106
        Purpose...106
        Format and Variation................................106
        Presentation 107
    Availability Indicators..107
        Purpose...107
        Format and Variation................................108
        Presentation 109
    Reliability Indicators...110
        Purpose...110
        Format and Variation................................110
        Presentation 112
    Maintainability Indicators 113
        Purpose...113
        Format and Variation................................113
        Presentation 114
    Utilization Indicators..115
        Purpose...115
        Format and Variation................................116
        Presentation 117
Process Performance Indicators.....................................117
    Labor Productivity Indicators.................................118
        Purpose...118
        Format and Variation................................118
        Presentation 119
    Nonproductive Downtime Indicators 119
        Purpose...119
        Format and Variation................................120
        Presentation 120

Customer Satisfaction Indicators 121
    Purpose...121
    Format and Variation 121
    Presentation 122
Operational Misses and Error Rates............................ 122
    Purpose.. 122
    Format and Variation 122
    Presentation 123
Cost Performance Indicators 124
  Cost Rates 124
    Purpose.. 124
    Format and Variation 124
    Presentation 125
  Cost Breakdown by Categories.................................125
    Purpose...125
    Format and Variation 126
    Presentation 126

Chapter 7 Computerized Maintenance Management Systems..... 127
Introduction 127
CMMS Objectives 128
CMMS Functions 132
  Equipment Module...132
  Work Order Module...132
  Preventive Maintenance Module...............................133
  Safety Module... 134
  Labor Module 134
  Inventory Module 134
  Financial Module...135
  Calendar Module 135
CMMS Features..136
  Accessibility and Security..136
  Communication and Notification...............................137
  Data Entry and Presentation.......................................138
  Integration 139
  Flexibility and Customization 140
CMMS Implementation..141

SECTION III   The Post-Maintenance Era

Chapter 8   The Systems View of the Equipment Management
Process 145

    Introduction 145
    Environmental Suprasystem..147
    Goals and Values Subsystem...149
    Structural Subsystem 152
    Technical Subsystem 154
    Psychosocial Subsystem...156
    Managerial Subsystem...157

Chapter 9   New Changes in the Post-Maintenance Era 163

    Introduction 163
    Equipment Management Objectives....................................163
        Functional-Level Objectives....................................164
        Job-Level Objectives 168
    Organizational Structure Changes 170
    The Platform Ownership Concept 177
    Employee Skill Requirements 180
    Computerized Equipment Management Systems...............185
    Work Environment Changes 189
    Management Changes 191
    Summarizing the Post-Maintenance Era............................194

Chapter 10 Transformation and Implementation 199

    Introduction 199
    Environmental Studies 200
    Managerial Preparedness...201
    Goal and Value Changes 202
    Psychosocial Changes 203
    Technical Changes 204
    Structural Changes.. 205

Glossary...209

References.. 213

Endnote 215

Index... 217

List of Figures

Figure 1.1 Equipment time and states..9

Figure 1.2 Equipment failure patterns 13

Figure 2.1 Equipment maintenance cost behavior....................................21

Figure 4.1 The relation between PM time and overall availability 46

Figure 4.2 An overview of the RCM implementation process 58

Figure 5.1 Headcount calculation worksheet example 71

Figure 5.2 An equipment expense budget worksheet example 76

Figure 5.3 A department budget summary worksheet example...........78

Figure 5.4 An equipment priority worksheet example template..........82

Figure 5.5 An example of escalation structure....................................... 84

Figure 5.6 A general process for maintenance training and development.. 86

Figure 5.7 A training and development direction decision model 87

Figure 5.8 A training assessment example form 89

Figure 5.9 A training matrix example template 90

Figure 8.1 Systems model adapted for equipment management.........146

Figure 8.2 Framework for analyzing equipment management change 147

Figure 8.3 Environmental factors that have an impact on equipment management...148

Figure 9.1 Equipment management process under the functional setup 171

Figure 9.2 Equipment management process under the platform setup 174

Figure 9.3 Platform engineering roles and responsibilities 179

Figure 10.1 Implementation process for migrating to the post-maintenance era 200

List of Tables

Table 2.1   History of Equipment Performance Management 27

Table 3.1   Major Components Required for Integrated Circuit Testing....32

Table 3.2   An Extension of Equipment Management History 42

Table 4.1   Different Types of PM Triggers 49

Table 4.2   Common PM Tasks and Examples..51

Table 5.1   Example Purchase Agreement... 96

Table 7.1   Common Business and CMMS Objectives............................129

Table 7.2   Typical CMMS Access Levels 137

Table 8.1   Summary of Environmental Trends in Equipment Management 150

Table 8.2   Summary of Objective Trends in Equipment Management ...151

Table 8.3   Summary of Structural Trends in Equipment Management 154

Table 8.4   Summary of Technical Trends in Equipment Management 156

Table 8.5   Summary of Psychosocial Trends in Equipment Management 158

Table 8.6   Summary of Managerial Trends in Equipment Management 160

Table 9.1   Key Knowledge Requirements of the Platform Owners 182

Table 9.2   Training Categories for Employees Involved in Equipment..184

Table 9.3   Comparisons between CMMS and CEMS 186

Table 9.4   Characteristics of Platform Ownership Concept..................195

Table 9.5   History of Equipment Management Summary......................196

Preface

In spring 2001, I took on the task of initiating a new course in equipment management for the master's in engineering management program at Santa Clara University. I searched for a book to be used for the course but found limited books that had a clear focus on equipment management. Most literature depicted how equipment should be managed under the field of maintenance management.

The maintenance management discipline was established in the 1950s with the concept of preventive maintenance (PM). It has evolved through the years and become a complicated field of study. Today, maintenance management includes key concepts such as maintenance prevention (MP), reliability-centered maintenance (RCM), predictive maintenance (PdM), and total productive maintenance (TPM). Most companies manage their equipment under the guidance of the principles in maintenance management.

Maintenance management and equipment management are not the same. Maintenance management is the discipline that concerns the maintenance function, which includes maintenance of other assets, such as building and facilities maintenance. Equipment management, on the other hand, specifically deals with equipment in manufacturing and engineering development. In many industries, equipment characteristics have changed significantly since the 1990s due to technology innovations. As a result, many maintenance principles are no longer effective for managing today's complex equipment. In fact, maintenance organizations along with the maintenance function have started to disappear in many high-tech companies, and in some cases, maintenance management can no longer be applied in equipment management. Perhaps it may even disappear or be replaced in the future. Whether or not this happens, the trend is certain that increasingly complex equipment is used in manufacturing and engineering development, and managing such equipment takes more effort, time, and money. Therefore, equipment receives increased attention, and a specialized field of study is desired.

Maintenance management principles are still suitable for traditional and stable manufacturing environments. However, they are outdated for complex high-tech equipment in the fast-changing business environments. For instance, one of the main objectives of maintenance management is to extend the life of equipment. The current fact is that, in many high-tech companies, equipment replacement rarely occurs because of the end of the natural life of the equipment; rather, replacement is driven by the introduction of new technologies and new products. The same phenomenon has been exhibited in the general consumer market; the main reason for people to replace their cell phones and computers is not that they are broken beyond repair but because of the new features and functions in the new generation of devices.

It is a trend for equipment to become computerized. Take automobiles, for example; they were originally pure mechanical devices that moved things and people from point A to point B. That is why auto repair technicians are called "mechanics." Today, almost all cars have computers controlling the fuel injection, brakes, turn stabilities, and so on. Can an old-school auto mechanic repair today's cars with just mechanical skills? No. With computerization, not only a computer is required but also sensors, often with advanced technologies in optics, temperature sensing, vibration detection, or oil chemical content recognition, as well as advanced interfaces and interconnections. Manufacturing is in the same trend of computerization, and computerized equipment will find its way into all manufacturing plants just as computers are going into almost every home in America. Eventually, those companies that do not upgrade their equipment will lose their competitiveness. History has shown us that what is considered "high tech" now will become the norm in the future, and based on the dynamic changes in the past several decades, that future is very soon. As equipment management practices depart from the maintenance principles, I believe that equipment management will provide guidance for managing equipment in manufacturing and engineering development settings, becoming an emerging discipline soon to be practiced by companies and with classes offered in many university programs.

I started my career as a maintenance professional in the mid-1980s. I found that a fundamental flaw was the maintenance functional setup.

As a functional department manager, I wanted my department to grow so I could survive in the corporate world. Growing the department means more employees and increased budget; however, all of this must be justified by the workload. As such, growing the maintenance business means more downtime, whether for PM or repairs. The company's output was impacted, and upper management was not happy. Conversely, if equipment performance was excellent and there was no downtime, the maintenance department may not exist, and I could lose my job. It is a dilemma that can only be addressed by moving away from the maintenance organizational setup. Therefore, for the past decade, my efforts have been to push maintenance organizations out of existence.

I wrote this book for three main purposes. First, it serves as a text for my students by covering the history and the fundamental knowledge of equipment management. Second, it shares my years of experience in managing equipment performance with other professionals in this field. Finally and most important, it presents new alternatives in equipment management beyond the current mainstream principles of maintenance management. These new alternatives are pioneered by such high-tech industries as semiconductor and biotechnology and are driven by the dynamic and fast-changing environment. As such, this book aims to initiate new thinking and approaches that will help organizations in the high-tech industries manage their ever-more-expansive equipment used in manufacturing and engineering development, as well as prepare companies in the traditional industries for the spreading of the microchip era in their equipment base.

Acknowledgments

I gratefully acknowledge all those who helped and supported me in completing this publication. Special thanks are due to Daniel Lamb for the opportunity to work in the leading environment of this field as well as the continuing support and mentoring for the past two decades and beyond. My sincere thanks also go to all the people who have worked alongside me—my peers and staff who offered essential support for my professional endeavors. In addition, I thank my teachers, professors, and students, who shared valuable knowledge through my long but enjoyable academic journey. Last, my deep appreciation goes to my relatives and friends, who have believed in my ability and offered words of encouragement throughout the years.

About the Author

Dr. Kern Peng holds two doctorate degrees, one in mechanical engineering specializing in nanocomposite materials and the other in business administration specializing in operations management. In addition, he also has an MBA in computer information systems and a BS in industrial engineering.

Dr. Peng designed and has been teaching the Equipment Management course at Santa Clara University, Santa Clara, California, since 2001. In addition, he regularly teaches four other master level courses in engineering management at SCU. Before that, he also taught MIS courses at San Jose State University, San Jose, California.

Dr. Peng has more than 26 years of people and project management experience in engineering and manufacturing, with over 19 years at Intel Corporation. He has mastered all aspects of engineering and manufacturing management and has proven results in finding innovative solutions to business and engineering problems. He has been accorded more than 50 career awards in the areas of engineering design; software development; technical paper publication; problem resolution; project management and execution; teamwork; and leadership.

Section I

Introduction to Equipment Management

1

Introduction

BACKGROUND

Today, most industries employ sophisticated equipment to produce products. Utilizing advanced equipment not only provides strategic advantages for many companies over their competitors, but also in many cases it is a necessity for survival. The recent developments in information systems and computer technology have led to an explosion of innovations in equipment and robotic technology. As a result, the characteristics of manufacturing equipment have changed significantly in recent years, especially in high-tech industries.

First, the complexity in equipment increases. For instance, in the past most machines were stand-alone units. Now, many machines are connected to a computerized manufacturing network to form an automated manufacturing line. Many machines have more parts and operate with higher speed and greater precision than previously.

Second, the fast-changing technology leads to a shorter equipment life cycle. In the past, the life of a machine generally ended when it failed and was beyond repair. Today, many machines are replaced due to technology obsolescence although they are still in good condition.

Finally, equipment cost has become increasingly expensive. For example, in the semiconductor industry, Rock's law was developed by venture capitalist Arthur Rock to signify this trend. Rock's law states that the cost of capital equipment doubles about every 4 years in the semiconductor industry, and this has been accurate since the beginning of the industry in the 1960s.

As equipment cost increases, the inefficiencies in equipment utilization have a greater impact on a company's output and profit. Managing

3

equipment performance has become a top priority in many companies. In the high-tech industries, the situation is made worse because the technology becomes obsolete at a rapid rate, leading to a short equipment useful life. To maximize returns on investment in a short period, companies often run their equipment longer and push the machines to operate at their limit in terms of speed and precision. As a result, managing equipment performance has increasingly become a challenging task.

Equipment management has gone through many phases. The first phase was the breakdown management phase in the pre-1950 period. In this phase, machines were only serviced when they needed repairs. The second phase was the preventive maintenance (PM) phase in the 1950s. During this period, the maintenance function was established, and time-based PM activities were generally accepted. The third phase was the productive maintenance phase in the 1960s. The key characteristics of this period were equipment reliability and maintainability focus, as well as cost consciousness in maintenance activities. The fourth phase was total productive maintenance (TPM) in 1970s. This period was significant for expanding maintenance techniques to include aspects such as human factors and equipment design. The fifth phase was TPM with predictive maintenance (PdM). This phase prevailed from the 1980s to the mid-1990s, when the condition-based maintenance concept called PdM was widely applied through the implementation of computerized maintenance management systems (CMMSs).

These phases are generalized by the acceptance of the particular equipment management practices by industries and firms in a specific period. This does not preclude that these practices have existed beyond the periods described. In fact, most of these practices are still in use in some industries and companies.

Equipment management has remained in the "maintenance" mode at large for decades. The maintenance methods are widely used and regarded as the norm in managing equipment. The conclusive focus of maintenance is on the optimization of equipment availability. However, only equipment utilization determines the output and profit of a factory. In most cases, factory organizational structure is set up in a way that the operations departments are responsible for equipment utilization, while the maintenance departments are responsible for equipment availability. Because of this functional separation, the whole picture is often not fully comprehended, allowing deficiency to occur in the overall equipment management process.

Because the traditional maintenance methods cannot meet today's requirements, especially in the high-tech industries, many leading high-tech companies have started to challenge the old norms in equipment management. From the organizational structure aspect, maintenance departments are disappearing. Maintenance functions are either absorbed in the operations via the universal tech concept or merged with equipment development via the platform ownership concept under new departments with titles such as manufacturing engineering, equipment engineering services, or platform engineering. The responsibilities of such departments extend beyond maintenance to include equipment utilization as well as equipment development.

The equipment platform ownership concept is usually practiced. The term *platform* is used to describe the inclusion of everything associated with equipment, from tangible elements such as network and facilities to intangible attributes such as software and network connections. In addition to keeping equipment operational, equipment platform owners are responsible for the entire equipment platform environment, from the development of equipment capability, platform support structures, methodologies, and comprehensive computerized equipment management systems to provision of a total solution to optimizing equipment output. The last part of this book defines this new phase in equipment management, called the "post-maintenance era."

This book is structured as follows: Part I consists of Chapters 1 to 3, and is an introduction to equipment management. Part II consists of Chapters 4 to 7; it discusses the maintenance system in detail. Part III, consisting of Chapters 8 to 10, proposes the new post-maintenance era and new methods in managing today's high-tech equipment.

Chapter 1 introduces the topic of equipment management and provides the key definitions of frequently used terminologies. Chapter 2 reveals the history of equipment management and defines five phases of equipment management in the pre-maintenance and maintenance era. Chapter 3 introduces the sixth phase, moving equipment management into the new post-maintenance era. Chapter 4 presents the general maintenance concepts. Chapter 5 describes the key logistics of maintenance management. Chapter 6 analyzes key maintenance performance indicators. Chapter 7 examines the CMMSs. The overall equipment management process in the systems viewpoint is exposed in Chapter 8. Chapter 9 explores the new changes in equipment management in the post-maintenance era. Chapter 10 demonstrates a step-by-step implementation plan that may be utilized to transform a company from "maintenance" oriented to the new platform ownership setup in the post-maintenance era.

MAINTENANCE MANAGEMENT

The dictionary definition of *maintenance* is "the work of keeping something in proper condition." In maintenance management literature, *maintenance* is generally defined as combinations of all technical and administrative actions intended to retain an asset or a system in, or restore it to, a state in which it can perform the required functions. In the daily vocabulary of traditional organizations, maintenance most likely refers to the department responsible for the function of maintenance. Obviously, maintenance is also considered a function of an organization. The objective of the function is to preserve either an asset or the ability of the asset to produce something safely and economically. Since maintenance does not produce any product, it is considered an overhead function.

Maintenance management is the discipline of maintenance. It originated in the 1950s with the implementation of the PM concept. Since then, it has been the mainstream discipline providing guidance for managing all company assets, but with manufacturing equipment as the center of attention. Most of the concepts and practices in maintenance management were developed from managing equipment and for managing equipment. Practices of other asset maintenance, such as building maintenance, remain at the early development stage of maintenance management and have not changed much. With managing equipment leading the development of the discipline, maintenance management has evolved to include key concepts and practices such as PM, maintenance prevention (MP), PdM, reliability-centered maintenance (RCM), and TPM. These concepts and practices are discussed in detail in Chapter 4.

Since maintenance is a function generally performed by the maintenance organization, it naturally comes with all the administrative logistics associated with a business unit. Maintenance management also provides guidance and strategies for managing these logistics. The maintenance administrative logistics include resource planning and budgeting, training and certification, customer services and management, vendor and contract management, and spare management, all of which are discussed in detail in Chapter 5.

EQUIPMENT MANAGEMENT

Ever since the establishment of the maintenance management discipline, maintenance management has been considered the principal guide for equipment management, and most of the key maintenance concepts and practices originated from managing equipment. In the minds of most people, equipment management is maintenance management. However, a fundamental change in equipment characteristics and business environments in recent years is that maintenance management principles are no longer providing effective guidance for managing equipment. This book makes a clear distinction between the two terms.

Maintenance management is more suitable for managing assets that are stable and consistent. Maintenance management principles do not yield the best results for managing equipment in such dynamic environments. For instance, one of the maintenance objectives is to extend equipment life. However, in many high-tech industries, equipment is obsolete due to technology changes much before the end of its natural life. As the importance of some of the maintenance objectives diminishes, the methods used in managing equipment must change as well.

The TPM movement started in the 1970s as an attempt to improve maintenance principles and to provide guidance for managing equipment in the increasingly modernized high-tech industries. However, it was focused on changing maintenance practices on the tactical level, and there were no fundamental changes in maintenance objectives and organizational structures. Maintenance continued to operate as a function of the organizations. The problem with the functional viewpoint is that it leads to the optimization of the function but not the overall process.

In the 1990s, reengineering started to become a popular concept for organizational changes. The reengineering concept focuses on improving the overall process by eliminating functions that do not make business sense or integrating functions that do not produce maximum value. As a result, the maintenance function started to disappear in leading high-tech companies. The foundation of the maintenance management discipline is weakening. It does not mean that maintenance management principles are obsolete.

Maintenance management is still suitable for a stable environment and stable equipment, such as building maintenance and machinery maintenance in traditional industries. However, a new discipline is needed to provide guidance for managing equipment in all areas. This book proposes the term *equipment management* as an emerging discipline of managing equipment in all aspects. It covers the maintenance management principles as well as the new concepts and practices beyond maintenance. Under the new discipline, equipment management is a process, not a function.

KEY EQUIPMENT TERMINOLOGY

Before going into detailed discussion of the topic, the key equipment terminology commonly used in the field must be clearly understood.

Equipment States

Equipment states are conditions of the equipment at a particular time. Generally, there are six basic equipment states: nonscheduled state, unscheduled down state, scheduled down state, engineering state, standby state (also called idle state), and productive state [1]. All other equipment states, such as shutdown state, setup state, and upgrade state, are categorized under these basic states. Each equipment state has a corresponding equipment time, which is shown in Figure 1.1. The definitions of equipment states are illustrated with the corresponding equipment time. Equipment states are determined by function, not by organization; as such, they are classified despite who performs the tasks, whether an operator, a maintenance technician, or an equipment engineer.

Equipment Time

Operations time is the period of time during which the equipment is expected to operate.

Nonscheduled time is the period of time during which the equipment is not scheduled to be utilized, such as nonworking holidays, weekends, and off-shifts. The situation is sometimes referred to as *shutdown*. Nonscheduled time includes shutdown time and startup time. *Shutdown time* is the time required to put the equipment in a safe condition when entering a nonscheduled state. *Startup time* is the time required for equipment to achieve

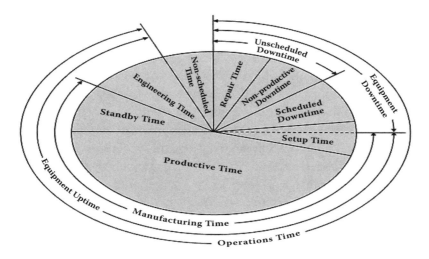

FIGURE 1.1
Equipment time and states.

a condition in which it can perform its intended function when leaving a nonscheduled state. Both shutdown and startup time are included in nonscheduled time. Nonscheduled time plus operations time should equal total time, which is at the rate of 24 hours a day and 7 days a week.

Equipment uptime is the period of time during which the equipment is in a condition to perform its intended function. It includes productive, standby, and engineering time and does not include any portion of nonscheduled time.

Manufacturing time is the period of time when the equipment is performing or available to perform its intended function. It is the sum of productive time and standby time.

Productive time is the period of time when the equipment is performing its intended function. It includes the time used for regular production, rework, and engineering runs done in conjunction with production on products that may or may not be production units. It consists of overhead time and product run time. *Overhead time* is the time spent in overhead events such as production data entry and product quality checks. *Product run time* is the time when the equipment is producing or testing the products.

Standby time is the period, other than nonscheduled time, when the equipment is in a condition and available to perform its intended function but is not operated. It means all supporting systems are available, such as chemicals, network, and facilities. It is also called *idle time*. The most common causes of standby are unavailability of operator or product, which brings up two terms: waiting operator (no operator) time and waiting work-in-progress (WIP) (no WIP) time.

Engineering time is the period when the equipment is in a condition to perform its intended function (no equipment or process problem exists) but is operated to conduct experiments in process engineering (e.g., process characterization) or equipment engineering (e.g., equipment evaluation).

Equipment downtime is the period when the equipment is not in a condition, or is not available, to perform its intended function. It includes scheduled and unscheduled downtime.

Scheduled downtime is the period when the equipment is not in a condition, or is not available, to perform its intended function due to planned events, such as PM, scheduled equipment specification verification, and equipment upgrade.

Unscheduled downtime is the period when the equipment is not in a condition, or is not available, to perform its intended function due to unplanned events, such as equipment failure or interruption of supporting systems (chemicals, network, and facilities).

Repair time is the period when the equipment is being actively restored to the condition at which it can perform its intended function. Repair time includes time needed for diagnosis, corrective action, equipment test, and verification run. *Diagnosis* is the action of identifying the source of an equipment problem or failure. An *equipment test* is the operation of the equipment to verify equipment functionality. A *verification run*, also referred to as a *standard run*, is a production run to establish that the equipment is performing within specifications.

Nonproductive downtime is the period when the equipment is not in a condition, or is not available, to perform its intended function, but no action is taken to bring the equipment to functionality. Sometimes it is referred to as maintenance delay. It may be caused by lack of awareness that the equipment is down or an administrative decision to leave the equipment down. It includes response time and waiting time for parts, a maintenance technician, equipment engineers, vendor field service engineers (FSEs), and so on. *Response time* is the time between the moment the equipment is down and the time when equipment support personnel respond to the incident.

Setup time is the period when the equipment is being prepared to perform its intended function. There is confusion regarding how to categorize setup time. To clear the confusion, two categories of setup, equipment setup and production setup, should be distinguished. *Equipment setup* includes equipment conversion, equipment test, and verification run (standard run). *Conversion* is the action taken to complete an equipment alternation to accommodate a change in process, product, package configuration, and so on. This type of setup time is generally categorized under equipment downtime. In some cases, depending on the operations and the companies, it is simply included in scheduled downtime. *Production setup* is the action to prepare for the production run. Examples of production setup tasks are product data entry, program loading, and sometimes a verification run. Production setup time is generally categorized under manufacturing time. Also, it is sometimes put under productive time and treated as part of the time required for normal production runs.

Equipment Actions

Breakdown maintenance is work performed in response to a breakdown. People commonly use the word *repair* for this situation, but there is a small distinction between breakdown maintenance and repair.

Repair is work performed in response to a failure. What is the difference between breakdown and failure? *Breakdown* is an unexpected interruption to the service of a particular asset. *Failure* is a termination of the ability of an asset to perform its required function at the standard level. For instance, in some cases, a breakdown occurs but there is no problem found after the investigation done by the maintenance personnel. In these cases, the asset does not have a failure; breakdown maintenance is performed but not repair. Breakdown maintenance is considered unscheduled downtime. Breakdown also includes an assist.

Assist is an unplanned interruption that occurs during equipment operation when all three of the following conditions apply: (1) The interrupted equipment operation is resumed through external intervention, by an operator or user, either human or host computer. (2) There is no replacement of parts other than specified consumables. (3) There is no further variation from specifications of equipment operation. An assist is generally less than 15 minutes. Equipment *interrupt* is any assist or failure.

Preventive maintenance is a series of scheduled or planned tasks designed to reduce the likelihood of equipment failure during operation. These tasks include cleaning, inspecting, calibrating, and running diagnostics on equipment and its components. The purpose is either to extend the life of equipment or to detect critical wear that will soon cause the equipment to fail. The PM schedule intervals may be based on time, equipment cycles, units produced, or equipment conditions.

Equipment installation is a series of tasks to install equipment at the equipment user's factory location. These tasks include layout, assembly, facility hookup, network connection, seismic protection, and so on.

Equipment acceptance is a series of tasks to ensure the equipment is meeting the design specifications and safety requirements. These tasks include safety buy off, extensive equipment burn-in, calibration, diagnostic runs, and verification runs with actual product units. Acceptance includes factory acceptance and site acceptance.

Factory acceptance is a series of equipment quality assurance tasks done at the equipment manufacturer's factory. The equipment manufacturer generally performs this process but sometimes this occurs with the equipment purchaser's representative present. The equipment manufacturer must demonstrate that the equipment can perform its intended functions within specifications defined by both parties.

Site acceptance is a series of acceptance tasks done at the user's installed site. Site acceptance is generally done jointly by the equipment manufacturer and the equipment purchaser. Site acceptance is also done after the equipment is transferred from one equipment user's factory site to another.

Equipment development is a series of tasks to further equipment capability, safety, ergonomics, reliability, and maintainability at the equipment user's factory location. These tasks include experimentation, prototype development, characterization, modification, upgrading, equipment testing, and so on.

Equipment Failure Patterns

By plotting the probability of failure against time of the equipment in commission, failure patterns can be revealed. Figure 1.2 demonstrates some of the failure patterns believed to be exhibited in equipment. Before

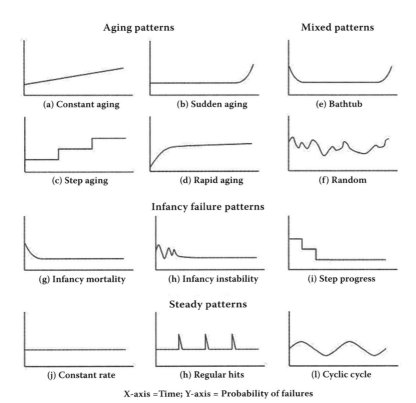

FIGURE 1.2
Equipment failure patterns.

the 1950s, the early view of failure patterns was simple: As equipment got older, failure was most likely to occur. These are the aging patterns (Figures 1.2a–1.2d). During the 1960s and the 1970s, the growing awareness of equipment "infancy mortality" led to the well-known "bathtub" failure curve (Figure 1.2e). In the 1990s, modern technologies began to be introduced in equipment, which led to some research findings showing that the failure pattern (Figure 1.2g) was the most common form of failure [2,3]. Progressing through the 2000s, as computer and mechatronics technologies matured, equipment changed significantly, which led to new generations of equipment with much more complex characteristics and failure patterns (Figures 1.2g–1.2l).

The basic belief relating equipment failure to old age has been challenged as the new generations of high-tech equipment address many of the failures related to operating age. It is now apparent that there is increasingly less connection between the likelihood of failure and equipment age. In addition, many of today's high-tech machines are obsolete due to technology or product generation renewal rather than age. The tail end of the bathtub curve and the correlation between failure and age become rather insignificant. While the infancy failure patterns are still common in complex equipment, steady failure patterns are often seen as equipment vendors are driven to fine-tune the equipment before sending it to customers to improve customer satisfaction. These increasingly common failure patterns indicate that once the infancy mortality issues are addressed, routine maintenance plays a minor role. This downplays the importance of maintenance and gives equipment management a whole new perspective. Many failure patterns are often seen on the same factory floor, and management needs to take appropriate actions based on the different failure patterns.

Equipment Performance Measurements

Safety is a combination of measurements defined by state and federal environmental, health, and safety (EHS) laws and standards. Such measurements include injury rates, lost-day and recordable cases, EHS violations, and potential safety incidents.

Availability is the percentage of time that the equipment was, or the probability that the equipment will be, in a condition to perform its intended function during a specific period. There is a difference between availability and uptime percentage. Availability calculations use manufacturing time as the numerator, and uptime calculations use equipment uptime, which is manufacturing time plus engineering time. Availability indicators include total availability, operational availability, equipment availability, operational uptime percentage, equipment uptime percentage, and so on.

Reliability is average interval that the equipment could, or the probability that the equipment will, perform its intended function within stated conditions. The average interval can be time, production unit, or equipment cycle. Reliability indicators include mean time between failures (MTBF), mean time between assists (MTBA), mean cycles between failures (MCBF), and so on.

Maintainability is the average time that the equipment was, or the probability that the equipment will be, retained in, or restored to, a condition to perform its intended function. Maintainability indicators includes mean time to repair (MTTR), mean time to assist (MTTA), mean time off line (MTOL), and so on.

Utilization is the percentage of time that the equipment was performing its intended function during a specific time. Utilization indicators include total utilization, operational utilization, engineering utilization, and others.

The details of these equipment measurements are presented in Chapter 6 on maintenance performance indicators.

2

History of Equipment Management

INTRODUCTION

For many years, machinery provided better alternatives compared to manual labor because machines could perform labor-intensive tasks faster and be more consistent. Today, machinery is no longer an alternative but rather a necessity in many industries. For instance, in the semiconductor industry, it is impossible to manufacture integrated circuits (ICs) without sophisticated machinery.

The increasing dependence on equipment prompted the desire to keep the equipment in good condition. As the use of equipment moved away from an alternative to become a necessity, the methods in managing equipment have also transformed. Until now, the history of equipment management can be categorized into five phases that cover the pre-maintenance era and the maintenance era. Equipment management has been largely operating under the maintenance management principles.

PHASE 1: BREAKDOWN MANAGEMENT

The breakdown management phase is the pre-1950s period when machines were mainly used to increase production output. The use of equipment was an alternative but not a necessity. In most cases, manual labors served as fail-safe backups. Therefore, equipment performance was generally not tracked. Management only paid attention to a machine when it was out of order. This can be characterized by a commonly heard phrase: "If it ain't broke, don't fix it."

The breakdown management phase is categorized under the pre-maintenance era. *Maintenance* is defined as combinations of all technical and administrative actions intended to retain an asset or a system in, or restore it to, a state in which it can perform the required functions. Before the 1950s, the maintenance function, as defined, was not established in organizations. The so-called maintenance department did not exist in firms. The focus on equipment was only to perform repairs when machines were down. Since equipment breakdown management was the sole action taken toward equipment management, the objective of this phase was to complete repairs in a reasonable amount of time that did not create a significant impact on production.

During the breakdown management period, managers often did not pay attention to equipment performance unless downtime gated production output, and they soon found out that they were firefighting equipment problems reactively, increasingly, and frequently. These unplanned events not only had an impact on production schedules but also caused product defects and in some cases led to customer complaints when the problems were not contained internally. The need to reduce unscheduled downtime and defects moved equipment management into a new phase: the preventive maintenance (PM) phase.

PHASE 2: PREVENTIVE MAINTENANCE

The first scientific approach to equipment management started in the 1950s with the application of the PM concept. During this period, PM was advocated as the sole means to reduce equipment failures and unplanned downtime. PM has been overwhelmingly accepted since the 1950s and is still one of the key practices. The PM concept set an important milestone of equipment management: the establishment of the maintenance function and the field of maintenance management. It was the beginning of the maintenance era that lasted until the appearance of the post-maintenance era, which is proposed by this book.

Preventive maintenance is defined as performing a series of scheduled or planned tasks that either extend the life of an asset or a system or detect critical wear that causes the asset or system to be about to fail [4]. As such, the objective of PM is to reduce unscheduled or unplanned equipment failures and to make equipment last longer.

The main benefits of reducing unscheduled failures are (a) to reduce product defects caused by equipment breakdowns and (b) to allow better capacity planning for equipment and product output. The benefit of extending equipment useful life is to reduce equipment acquisition costs under the assumption that equipment does not become obsolete when new generations of products are introduced. Overall, PM was a significant development that allowed managers to have some control over equipment failures. In the previous breakdown management phase, equipment failures were seen as acts of nature, and one could only deal with them as they occurred.

During the 1950s, PM was limited to time-based maintenance programs. In many companies, PM schedules were implemented on all major equipment. Lists of checkout tasks were planned and performed weekly, monthly, quarterly, semiannually, and/or annually. The routine PM practice quickly became a popular method in equipment management. Companies set up maintenance departments that consisted of dedicated personnel for performing the PM tasks and equipment repairs.

Equipment manufacturers also caught up with the trend. They believed that performing PM could reduce their cost of after-sale services, especially during the warranty period. Consequently, equipment manufacturers began to ship equipment with a recommended list of routine checkout items to be performed on a regular schedule. Companies that acquired the equipment had to perform the PM tasks based on the schedule to keep the warranty valid. This concept is still used today and is one of the most popular practices in equipment management.

Although PM has its benefits, it creates inefficiency in equipment utilization. Because equipment manufacturers required it, performing PM became a habit. However, most PM tasks require equipment to be taken off line for inspection. As such, PM is categorized as scheduled downtime, and all downtime affects product output.

Managers often accept PM downtime on the assumption that it represents an investment against future downtime incidents, and they seldom ask these questions: What determines the current PM schedule? Can PM be done less often? Or more aggressively, how can PM be eliminated by improving equipment reliability or by setting up a support system that can foresee future equipment failures? Some maintenance literature stated that PM was a means to ensure continuous and smooth operation of equipment. This view is incorrect because PM itself, as a scheduled downtime event, discontinues operation in most cases. In addition, not all PM tasks yield benefits against unscheduled downtime. There was a need for optimization, which became the focus of the next phase in equipment management.

PHASE 3: PRODUCTIVE MAINTENANCE

Since general applications of reliability theory were proven successful in Japan, many firms started to apply statistics to equipment management, which was a step into the productive maintenance phase. Operations research models for maintenance first appeared in the 1960s. Reliability, maintainability, and maintenance efficiency became the focus. Statistical indicators, such as mean time between failure (MTBF) and mean time to repair (MTTR), were used to measure equipment reliability and maintainability.

In addition to reducing unscheduled events and defects, the objective of the productive maintenance phase was to increase maintenance efficiency. During this period, three important concepts were introduced to equipment management and are still widely applied today. First, the concept of reliability reveals how reliable or dependable equipment is. It is measured by the indicator MTBF, which is an average of how often a machine fails. For instance, a machine with a MTBF of 500 hours means that this particular machine is likely to fail every 500 hours on average. Obviously, the higher the MTBF, the more reliable the machine is. As an indication of the frequency of equipment failure, MTBF is an important indicator for determining PM schedules.

Second, the concept of maintainability illustrates how easily equipment can be maintained or repaired. It is measured by the indicator MTTR, which is an average repair time of the failures. For example, a machine with an MTTR of 5 hours means that it takes an average of 5 hours to repair the machine each time it goes down. Clearly, the lower the MTTR, the easier the machine is to maintain. As a calculation of average repair time, MTTR can also be used to track maintenance efficiency, skills of maintenance personnel, and training effectiveness.

Third, the concept of engineering economy was applied to equipment management in this period. During the preceding PM period, large time-based PM programs were widely implemented, and maintenance departments were set up in many corporations. As the maintenance function was established in organizations, the costs incurred in maintenance became an important line item in operating expenses. During the productive maintenance phase, companies became cost conscious regarding maintenance activities. Management started to develop maintenance budget plans, track maintenance expenses, and reduce maintenance costs.

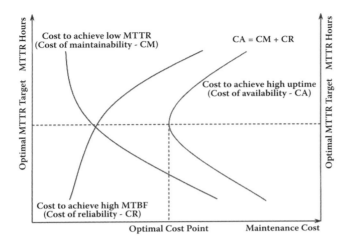

FIGURE 2.1
Equipment maintenance cost behavior.

Combinations of MTBF and MTTR were used to predict the costs of repairs and unscheduled work. Figure 2.1 demonstrates the cost behavior of equipment maintenance. The costs of planned maintenance or scheduled work were determined by the PM frequency and average time needed.

In addition to maintenance costs, reliability and maintainability, as measured by MTBF and MTTR, respectively, also created a significant impact on equipment acquisition and operations capacity planning. Using the equipment reliability and maintainability indicators, planning staff could statistically determine the number of machines required to produce certain output with a certain degree of confidence. These indicators allowed companies to plan operations and measure the execution based on the plans.

This approach might seem proactive because it predicted equipment failures and accommodated downtime into planning models. However, it was still considered reactive in equipment management because management implicitly accepted the fact that equipment downtime would occur. The mentality was that the operations would run under control as long as equipment downtime was calculated and put into the capacity-planning model. So, when equipment did go down, it became no surprise because it was expected. When the desired output could not be achieved, the capacity-planning model was blamed for being inaccurate, and more machines should be purchased. This method is still popular today within many firms even in the high-tech industries.

Another concept worth mentioning is reliability-centered maintenance (RCM), which was also founded in the 1960s but initially was oriented toward maintaining airplanes and was used by aircraft manufacturers, airlines, and the government. Not until two decades later did it start to spread to other industries. RCM is an analysis method that directs maintenance efforts at those parts and units for which reliability is critical (i.e., safety).

There are five steps in the RCM approach. First, all functions of the equipment are identified [4]. Second, all ways that the equipment could lose functionality are examined. Third, each loss of function is reviewed to determine all of the failure modes that could cause the loss. Fourth, the consequences of each failure mode are analyzed and put into four categories: safety, environmental damage, operational, and nonoperational. Finally, action plans are developed to detect the before-fail conditions to avoid failures; otherwise, fail-safe systems are implemented to avoid the consequences when failures occur. Such action plans and systems should be technically possible and make sense.

PHASE 4: TOTAL PRODUCTIVE MAINTENANCE

Started by the Japanese automotive industry, particularly Toyota and its related companies, equipment management moved into a new phase by using the concept of total productive management (TPM). TPM is a concept that combines the American practice of PM with Japanese total quality control (TQC) and total employee involvement. The approaches in the PM and productive maintenance phases were concentrated on equipment maintenance techniques. The focus of both phases was on the technical aspect of equipment management, which fell under the umbrella of scientific management principles developed by F. W. Taylor. Following the human relations movement in general organizational management, equipment management also moved from its sole technical focus to include human factors as well.

Japanese firms started to practice TPM in the 1970s, but American companies did not catch on until the late 1980s, when two of Seiichi Nakajima's books, *Introduction to TPM* [5] and *TPM Development Program* [6], became available in English. However, TPM gained much attention just shortly after its introduction in the United States, and American manufacturing firms overwhelmingly accepted it.

According to Nakajima, the dual objectives of TPM are zero breakdowns and zero defects. To achieve such aggressive objectives, TPM emphasizes the importance of operator involvement in making equipment reliable. The assumption is that caring about the job cannot be taught, and TPM can create an environment that encourages employee commitment to meet the goals [5].

TPM is a result of the implementation of total quality management (TQM) and just in time (JIT). In fact, TPM should be viewed as a TQM extension applied to the field of maintenance management, and JIT forced the implementation of TPM because JIT would not be possible if equipment had frequent breakdowns. Dr. Tokutaro Suzuki, senior executive vice president of Japan Institute of Plant Maintenance, said, "Total productive maintenance (TPM) is indispensable to sustain just-in-time operations" [4, p. 135]. Hence, Japanese firms implemented all three concepts as a package to enhance overall factory performance.

The TPM implementation during the 1970s could be grouped into three major components: maintenance prevention, PM, and autonomous maintenance. The first component, maintenance prevention, is defined as an approach to eliminate equipment breakdown through designing or selecting equipment that is maintenance free. Under TPM, maintenance was no longer the sole concern of maintenance departments and maintenance personnel. By using the maintenance prevention concept, equipment maintenance is considered upstream in the equipment design and selection phases. Obviously, involvement of equipment vendors and suppliers is required in the same way as in TQM and JIT implementations. Equipment designers are taking steps to keep equipment design simple and to eliminate parts that require frequent maintenance. In addition to capability and cost, equipment selection is generally done with the inputs from maintenance personnel based on the ease of maintenance as a criterion.

Preventive maintenance, an American approach, is still a key component of TPM. However, some basic maintenance tasks are passed down to machine operators. Under TPM, maintenance and production have a close alignment. The responsibilities of the maintenance department are to perform major repairs and PM, train operators, set standards, and consult on maintenance improvement ideas.

Last but not least, autonomous maintenance is the most essential component and the backbone of TPM. Autonomous maintenance is a strategy that involves production personnel in the total equipment maintenance process. Hence, TPM involves not only equipment vendors and suppliers upstream but also equipment operators downstream. Autonomous maintenance empowers equipment operators to make decisions about equipment performance and gives them the opportunities to increase their job skills. Therefore, it is the production department's answer to employee empowerment, job enhancement, and total quality improvement programs. It also breaks down the departmental barrier that exists between maintenance and production.

Before TPM, operators would leave the machine once it was down, and the responsibility would then be passed on to the maintenance personnel. Operators simply did not care and showed the "it-is-not-my-job" syndrome. There were cases where operators put equipment down so that they could go for a break. When the production goals were not met, equipment downtime was often the excuse. Maintenance personnel, on the other hand, blamed the operators as incompetent and troublemakers. Autonomous maintenance eliminates the tug of war between maintenance and production departments and allows them to work as a team.

TPM has been recognized as a significant improvement in the history of maintenance management. It has the ultimate goal of zero breakdown and defects. However, these two attributes are mainly associated with unscheduled downtime and events. The major disadvantage of this approach is that it hurts the overall availability by taking equipment time regularly and nonoptimally to achieve such an aggressive goal. TPM has the standards of 90% availability, 95% performance efficiency, and 99% rate of quality parts [4]. Supposedly, all these TPM standards are achieved in a given factory; the equipment in this factory would still be unavailable for 10% of the time. In 7-day, 24-hour operation, equipment would be off line for 16.8 hours per week or approximately 2.5 hours per day for routine PM services and inspections. In the majority of cases, this does not meet today's equipment performance standards, especially in the semiconductor industry, for which an equipment availability goal generally ranges from 95% to 98%.

Another maintenance management development that is worth mentioning is terotechnology, which is a term originated by Dennis Parkes in the United Kingdom during the same period of TPM implementation in Japan. The definition of *terotechnology*, according to the British Standards Institute, is

> a combination of management, financial, engineering, and other practices applied to physical assets in pursuit of economic life-cycle (LCC). Its practice is concerned with the specification and design for reliability and maintainability

of plant machinery, equipment, buildings, and structures, with their installation, commissioning, maintenance, modification, and replacement, and with feedback of information on design, performance, and costs. [5, p. 18]

PHASE 5: TPM WITH PREDICTIVE MAINTENANCE

Since the 1980s, computer technologies have been increasingly applied to businesses. The use of computers enabled maintenance managers to optimize maintenance activities further. Computerized maintenance management systems (CMMSs) were widely implemented to enhance the effectiveness of equipment management. While TPM continued to gain acceptance in the United States, the concept of predictive maintenance, which is a condition-based maintenance method, was added to maintenance management as a new component of TPM. The focus of this phase is on optimizing both unscheduled and scheduled events to maximize total equipment availability.

Predictive maintenance focuses on determining the life expectancy of components to replace them or service them at the optimum time. The concept of predictive maintenance has been developed for decades; however, implementing such a concept had been mostly infeasible until the advanced computer-aided design (CAD) and computer-aided manufacturing (CAM) technologies became available.

The utilization of computer technology has transformed equipment to "smart" machines. It is common for equipment with computerized control to have the capability for self-monitoring, self-calibrating, and self-adjusting. The development of advanced technologies in other areas, such as infrared, vibration analysis, noise and optical sensing, also helped the implementation of the predictive maintenance concept. By monitoring the before-fail symptoms of machines, maintenance services can be performed at the right time on the right components. Hence, equipment availability as well as maintenance labor hours are optimized.

In addition to improving equipment capability, computer technologies were applied to keep track of maintenance activities, costs, and equipment failures. CMMSs have been the central focus for equipment management since the 1980s, and they have been viewed as keys to achieving maintenance efficiency.

CMMSs started as simple database software programs that kept records of maintenance work orders or down tags. Such database programs significantly increased the speed of gathering and cross-referencing equipment information, such as frequency and average repair time of a particular failure, average PM time of certain maintenance personnel, and the like. The local-area networking (LAN), wide-area networking (WAN), and office computing technology allowed CMMSs to be accessed locally or remotely, which made information sharing easier, especially for companies that had multiple factories located all over the world.

Artificial intelligence (AI) was another significant development during this phase of equipment management. Based on the AI concepts, expert systems were developed to enhance the equipment maintenance skills of maintenance personnel. These expert systems guided maintenance personnel in performing repairs and also served as training tools. Some expert systems were developed as stand-alone software programs, but others were included in the CMMS so that the intelligence continued to build as new data were collected by the CMMS.

In this phase of equipment management, the application of computer technologies is the key characteristic. Many previously developed maintenance concepts are brought to a new optimal level of applications because of the advancement of computer technologies. Fundamentally, the approaches to equipment management are still under the guidance of the maintenance theories developed in the previous phases.

SUMMARY OF THE PRE-MAINTENANCE AND MAINTENANCE PHASES

In this discussion, equipment management has gone through the five phases summarized in Table 2.1. The general time period and the characteristics of each phase are presented along with the equipment management objectives and concepts developed.

Phase 1 is considered the pre-maintenance era, and phases 2 to 5 are the maintenance era. These phases were categorized into specific periods based on the general acceptance of such practices by pioneer industries. Many of these practices have existed beyond the described periods. In fact, most of these practices are still in use in some industries and companies. Multiple practices across from multiple phases can also be seen in a single company depending on the complexity of the equipment, from simple devices to complex computer-controlled robots.

TABLE 2.1

History of Equipment Performance Management

	Period	Characteristics	Objectives	Concepts Developed
Phase 1 Breakdown management	Pre-1950	Repair only when machines were down	Repair equipment failures in reasonable time	"If it ain't broke, don't fix it."
Phase 2 Preventive maintenance	1950s	Establish maintenance functions Time-based maintenance	Extend equipment life Reduce unscheduled downtime and defects	Preventive maintenance Productive maintenance Maintainability improvement
Phase 3 Productive maintenance	1960s	Reliability focus Maintainability focus Cost conscious	Reduce unscheduled downtime and defects while increasing maintenance efficiency	Reliability engineering Maintainability engineering Engineering economy Reliability-centered maintenance (RCM)
Phase 4 Total productive maintenance (TPM)	1970s	Preventive maintenance plus TQC and total employee involvement	Zero breakdowns and zero defects	Behavioral sciences Systems engineering Ecology Maintenance prevention Just in time (JIT) TQC and TQM Terotechnology
Phase 5 TPM with predictive maintenance	1980s–1990s	TPM practices Condition-based maintenance Application of CMMS	Zero breakdowns and zero defects Optimization of availability	Computerized maintenance management Artificial intelligence and expert systems

3

Introduction to the Post-Maintenance Era

In many high-tech industries, especially in the semiconductor industry, business environments change at a fast pace. and rapid changes in equipment technologies and characteristics follow. Equipment management faces new challenges related to complexity, expensive costs, short life cycle, and extensive use of equipment. The traditional maintenance approaches do not seem to deal with these issues effectively. Ever since the late 1990s, the semiconductor industry has started a new trend that moves the field of equipment management out of the maintenance era. Before discussing the post-maintenance era, the following sections in this chapter use the semiconductor industry as the primary example to examine the business environment changes that led to the movement of transitioning out of the maintenance era.

THE NEW BUSINESS ENVIRONMENT

In recent years, there have been significant increases in production process changes as well as the rates of product introduction and obsolescence. These operational changes have led to significant changes in equipment characteristics. Equipment has become increasingly complex. In addition, frequent equipment changes are required to meet the needs of frequent operational changes. Equipment useful life is reduced due to rapid rates of product and technology obsolescence. As a result, costs associated with equipment management have significantly increased, as reflected in Rock's law in the semiconductor industry.

The new business environment in the high-tech industries, especially in semiconductor manufacturing, is far different from the traditional manufacturing environment in which the maintenance function originated. Such environmental changes have prompted the need for new equipment management approaches. New technology and management concepts that were developed during the past decades have set a foundation for the development of the new equipment management approaches.

Operational Changes

The need for equipment is driven by production requirements. As the operational scope changes, the way equipment is used and managed changes as well. In many industries, manufacturing processes have changed more frequently in recent years. Again, using the semiconductor industry as an example, the early manufacturing process technology was under the "micron" naming convention and has changed to "nanometer" as the processes improved. The current 32-nanometer process technology means that the circuit lines on a chip have a width of 32 nanometers, about 4,000 times smaller than a human hair.

Long before the process technology reached below the submicron level, many industry analysts speculated that semiconductor manufacturing was close to the limits imposed by physics. The majority of the limitations are associated with equipment as the difficulty factors increased substantially with each new generation of process implementation. Each process technology change represents a significant change in equipment. Yet, the semiconductor process technology continues to improve, at even a faster rate. Such achievements are largely the result of overcoming major obstacles in equipment technology.

Not only the manufacturing process changes but also new product introduction affects many aspects of equipment design and management. Within the same semiconductor manufacturing process, various products often have different equipment requirements. Each new product, from development to manufacturing, whether it is a new generation or capability enhancement of a previous version, often requires equipment changes.

As new products continue to enter the market at a rapid rate, older products are rendered obsolete. The product introduction and obsolescence rates have increased considerably in the microprocessor segment since the beginning of the twenty-first century. The microprocessor segment is considered the driving force of the industry, and it drives other segments, such as chipsets and memory products, to keep up with such rapid changes. Therefore, it best represents the overall semiconductor industry dynamics. During the maintenance era, Intel introduced about 1.5 new microprocessors per year on average, while since 2000, the average rate of new microprocessors introduced each year quadrupled.

Clearly, the operational changes are phenomenal in the semiconductor industry. The semiconductor industry is generally considered the backbone of other high-tech industries. The new products produced by the semiconductor industry have led to many new technology implementations in other industries. The product introduction and obsolescence rates have increased significantly since 2000 in sectors such as the computer industry and the consumer-electronic industry. As a result, there is a much faster pace of operational changes in the overall manufacturing sector.

Equipment Characteristics

The operational changes mentioned affect equipment characteristics significantly. Using the semiconductor industry as an example again, many industry analysts believed that the semiconductor manufacturing technology was very close to its physical limits. There were many obstacles to and difficulties in having a stable process to produce high-speed integrated circuits (ICs). However, the industry has been continuously moving forward in process generations by increasing equipment capabilities to cope with environmental variations.

Each implementation of process technology since the year 2000 represented a significant breakthrough in equipment technology. Equipment capability has been keeping up with new product introductions. For instance, the equipment used to test today's microprocessors has a significant capability improvement over the testing equipment used to test the 386 and 486 microprocessors.

However, such improvements accompanied the ever-increasing complexity of equipment. For example, the first generation of microprocessor testers was merely a computer with enhanced performance. Today, microprocessor testers have complex and sophisticated electrical, mechanical, and optical components with several thousand parts and may take up a small room. It is typical for a tester to carry a multimillion-dollar price tag.

Testing the IC chips becomes complex and requires many additional components, as summarized in Table 3.1. Any problem with one component affects product quality and output. In addition, test equipment must

TABLE 3.1

Major Components Required for Integrated Circuit Testing

Major Components	Purposes of the Components
Circuit breakers, power supplies, power stabilizers, and uninterrupted power supply	Ensure clean and uninterrupted power; the higher the frequency of the product, the cleaner power it requires.
Circuit boards, high-speed cables, connectors, and relays	Provide analogue and digital testing capability; these components increase in number as new chips have more pins.
Heat exchangers, valves, piping and plumbing for chill water, coolants	Provide cooling to circuit boards; the higher the frequency of the product, the more heat it generates.
Mechanical parts, valves and piping for compressed air and vacuum	Allow test head movement to dock on automatic wafer or chip handling equipment.
Automatic wafer probers and chip handlers	Provide automated wafer and chip handling for high-volume manufacturing.
Test interface units (TIUs) and probe card	Provide interface with tester and prober or handler.
Vision sensors and interlocks	Provide safety for accessing hazardous zones.
Thermal devices, valves and piping for liquid nitrogen and nitrogen	Provide thermal control so products can be tested at various temperatures.
Computer workstations and equipment control software	Provide control and diagnostic capability to all functions of the equipment and network connectivity.
Network connection devices, phone lines, LAN, servers, and product programs	Provide network connectivity to access product programs, which are product specific and stored in central servers.

be operated under a controlled environment dictated by the required clean room specifications. Room temperature and humidity changes can affect the products as well. Therefore, semiconductor manufacturing is considered one of the most complicated processes in manufacturing, and its level of complexity is increasing with each new generation of process technology and products.

With frequent manufacturing process changes and new product introduction, changes are made to equipment at an unprecedented rate. Research documentation indicated that each process technology changeover affects over 80% of the equipment base and leads to a 10–20% increase in equipment number. New equipment must be installed or existing equipment must be upgraded to meet the new process specifications. New products often require equipment upgrades as well. Therefore, equipment changes become routine tasks of equipment management, averaging once to twice a week in any given facility.

In addition to the frequent changes imposed on equipment, the short life cycle of process technology and products results in a shorter equipment useful life. In semiconductor manufacturing, the majority of equipment becomes obsolete because the newer generation of process technology and products has higher specification requirements, not because of the natural wear and tear of the equipment. As a result, the failure patterns are most likely not age related, as shown in the equipment failure patterns section in Chapter 1. It means that routine maintenance has a minimal effect on the probability of failure. Therefore, the traditional maintenance tactics may no longer be effective.

Undoubtedly, the complexity and rapid changes in equipment impose significant cost increases. The costs of capital equipment and machinery are very high and are increasing with each generation of process technology. Semiconductor manufacturing equipment has been getting increasingly expensive each year. In short, as Rock's law states, equipment acquisition cost doubles every 4 years. In addition to acquisition cost, which is only a part of equipment costs, equipment installation cost increases as new equipment is installed frequently. Installation of new equipment may also include facilities changes such as building a new clean room and new power stations, upgrading chill water and compressed air capacities, or acquiring new network servers. Because of the complexity, spare part inventory and training requirements have increased, leading to an increase in overall maintenance costs.

Again, the semiconductor industry is the best demonstration of the recent changes in equipment characteristics, but other industries have experienced the same trend. In general, equipment has become increasingly complex and expansive. Equipment is also used extensively to achieve higher profits.

New Enabling Technologies

In the years since 1990, we have witnessed many technological innovations. These innovations not only enhanced equipment capabilities but also modernized equipment management. Before the mid-1990s, in semiconductor manufacturing most machines were controlled by computers using the UNIX operating system (OS), which is technically oriented toward engineering and hardware interfaces. The UNIX workstations were the preferred choice for technical personnel as their main computing tools. On the other hand, managers and other equipment management staff, such as planning and purchasing personnel, used personal computers (PCs) operating in the Microsoft® Windows environments, which provide better office software for planning, scheduling, and business presentations. The use of different computing tools created a communication gap between different parties involved in equipment management. Several recent technology developments have significantly reduced this gap, and the one with the most impact is the development of the Internet and intranet.

Using HTML (hypertext markup language), DHTML (dynamic HTML), XML (extensible markup language), and other applications programs, such as Java and ASP, the Internet and intranet became independent of the OS interfaces, significantly enhancing communication among platforms that use different OSs. Previously, equipment performance data captured in the UNIX workstations often had to be transferred to the PC platform using standard text files and then extracted by Microsoft Excel or other PC-based programs for reporting and graphing. It was also difficult to write object-oriented application programs that ran on both UNIX workstations and PCs. With Internet and intranet technology, information can be displayed or edited through Web pages regardless of the OS platform where the data is stored. Today, many computerized equipment management systems have a Web interface to allow data entry and access everywhere in the company, on the manufacturing floor with an equipment control workstation or in the office with a PC.

In addition to the Web technology, many other developments have contributed to closing the gap between different OS platforms. Starting with Windows NT, Microsoft addressed the issue by enhancing the capability of the OSs to provide a better network connectivity and interface with other platforms as well as better hardware handling. The Linux OS is another recent development that attempts to resolve multiplatform issues. Linux is also platform independent and is free for everyone who wishes to use it. As a result, many equipment manufacturers started to migrate the equipment OS from UNIX to a Windows-based OS or Linux.

Besides the new technology developments in computing, telecommunication technologies have advanced, and many have been applied to equipment management. For instance, text messaging is used for equipment status notification. When a machine goes down, technicians will receive a message on their pagers or cell phones showing which machine is down and for what reason it is down. The advancement of wireless communication technology connects equipment personnel with their equipment anytime and anywhere. The semiconductor industry is certainly on the leading edge of recent technology development. It is the best representation of the new economy and business environment, leading equipment management into a new chapter. Other manufacturing industries will certainly follow a similar path of development.

New Management Concepts

New management concepts also have a significant impact on the recent changes of equipment management practices. The most noticeable management concept in the 1990s was reengineering. Since the 1990s, the concept of reengineering has been applied to many areas of businesses, but in the field of equipment management, only minimal reengineering activities have been performed, and the majority of these activities were still conducted within the maintenance box. Maintenance has been viewed as a functional unit. The problem with the functional viewpoint is that it leads to the optimization of the function but not the overall process. The reengineering approach focuses on improving the overall process by eliminating functions that do not make any business sense or integrating functions that do not produce maximum value. In many cases, it results in elimination of tasks and functions and change of processes.

For example, equipment performance reports were generally done by administrative personnel. These employees, who devoted hours of work to generating the reports, were not always the ones who needed to do something with the data. Many people hated generating reports so much that they often did not look at the result and did not catch obvious contradictions. As a result, the reports were usually late and often contained errors. So, these people who generated the reports took actions to automate them. However, the problem still existed because only the tasks were improved. From the viewpoint of the employees who generated these reports, the problems were solved. However, the reports still contained errors, and the formats were difficult to understand. After applying reengineering, the report generation tasks were given to the equipment owners who had vested interests in the data. The reports were generated as needed with formats that fit different audiences.

When reengineering concepts were applied at the macrolevel of equipment management, it became clear that equipment maintenance had been viewed as a function since its establishment in the 1950s.

Whenever a new plant or factory was planned, the maintenance department was automatically put into the organizational chart without any objections or even questions for its necessity. Once the department was set up, any performance issues related to equipment would be assigned to the department; hence any improvements made were generally confined to the maintenance function. This functional focus led to the optimization of the function but not the overall manufacturing process. The reengineering approach focused on improving the overall process rather than specific functions. Therefore, it fostered the reengineering of the entire equipment management process beyond the traditional scope of maintenance.

In the late 1990s, another new management concept was proposed by Tom Peters in his book *The Circle of Innovation*. In this book, Peters analyzed many success stories of leading high-tech companies and concluded that the traditional focus on continuous improvements was too incremental and slow [7]. Hence, it could no longer lead to major successes. Rather, companies should focus on seeking ways to leap to a higher performance level. In many cases, this means a complete change in processes and work methods. The key elements that allow companies to leap in their performance are out-of-the-box thinking and paradigm shifts. Peters's concepts helped initiate further changes in equipment management practices. Leading companies started to conceptualize the issues of "maintenance" as larger than those contained in the traditional maintenance function or departments; this in turn led to the diminished role of maintenance departments in leading companies. Peters's concepts further initiated the changes in equipment management practices to a new level.

After several decades of optimizing the maintenance function, it is time to examine the overall equipment management process from outside the maintenance box by applying reengineering and Peters's innovation approaches.

THE ISSUES OF MAINTENANCE

Until phase 5, equipment management had been operating in the maintenance mentality at large. The theories in maintenance management do not cover the entire scope of equipment management as it prevails today, especially in the semiconductor industry. Although many maintenance management methods are still widely used today and are regarded as the norm in managing equipment, there are many issues, ranging from objectives to organizational setup.

Problems in Objectives

The first issue with maintenance management approaches is in the objectives. For instance, one of the key objectives in maintenance management is to extend the life of equipment. The importance of this objective has diminished in many industries because of the rapid rate of technology obsolescence. The replacement of equipment rarely occurs because of the end of the natural life of the equipment; rather, replacement is driven by the introduction of new technologies and new products.

Although total productive maintenance (TPM) had been a significant development that revolutionized the field of equipment management, it has its incompleteness because of its limited focus on reducing breakdowns and defects. The goals of equipment management, in today's terms, extend beyond restraining unscheduled events and scheduled preventive maintenance (PM) to include optimizing other equipment events, such as developments and upgrades. Hardware and software upgrades are done often on equipment in high-tech industries due to frequent technology changes and new product introductions. Despite the goal differences, some TPM practices can still be applied to equipment management in the semiconductor industry. However, new practices must be added to cope with the unattended issues.

Furthermore, maintenance management has a conclusive goal of optimizing equipment availability. However, it is equipment utilization that determines the output and profit of a manufacturing facility. For instance, if a machine is only utilized 5% of the time, extensive efforts in achieving 100% availability may not be necessary. Factory managers generally do not care about availability when there is a huge gap between utilization and availability. As the objectives change in the field of equipment management, the approach must be changed to obtain optimal results.

The objectives of the maintenance units are also in contradiction with the overall objectives of the factory. For instance, in most business units, the amount of work generally determines the headcount. The main tasks for the maintenance department are repairs and PM, which are categorized as unscheduled and scheduled downtime, respectively. To increase output and equipment utilization, both categories of downtime must be reduced. However, it means less work and less headcount for the maintenance unit.

Logically, it is not in the best interest of maintenance managers to reduce their headcounts.

Individually, a maintenance worker must compete for a reasonably large workload to become a top performer in the organization. The more work for the maintenance worker, the more downtime there is in equipment and the less output for the factory. This creates an issue in job performance evaluation because a well-performing maintenance worker is a result of equipment trouble and poor factory performance. Obviously, the maintenance organizational setup causes conflicts among personal, departmental, and corporate objectives. It is ineffective and must be changed.

Structural Inefficiency

The second issue of maintenance management is in its organizational structure. Under maintenance management approaches, maintenance is a separate functional unit. Factory organizations are set up in a way that operations or production departments are responsible for equipment utilization, while maintenance departments are responsible for equipment availability. In addition, other engineering departments are responsible for equipment selection, development, and upgrades. Inventory control or logistics departments manage equipment parts. Purchasing and legal departments negotiate equipment support contracts. The introduction of CMMS (computerized maintenance management system) added another party, the information technology (IT) departments, into the picture.

Although TPM makes the effort to involve other parties, such as equipment suppliers and operators, in the maintenance function, clear ownership of equipment management does not exist. The parties involved have their own functions and objectives that often make equipment performance management a lower priority. Conflicts exist among these departments, and the whole picture is often not fully comprehended, allowing deficiencies to occur in equipment management.

Managing equipment is not the highest priority of the other departments, making the objectives of the maintenance department difficult to achieve since the department generally does not have total control over equipment. For instance, it does not have control over what and when upgrades are performed because the nature and schedule of the upgrades are often determined by product manufacturing requirements and product introduction schedules.

Also, the maintenance department is generally viewed as an overhead function that does not add direct values to products. As a result, the maintenance budget is consistently scrutinized. For example, according to W. W. Cato and R. K. Mobley in their book *Computer-Managed Maintenance Systems in Process Plants* [8], the most difficult of all tasks in implementing a CMMS is selling and justifying the program to management. Seemingly, maintenance departments are given the responsibility of equipment management, but they are doing it with their hands tied. All of these issues indicate that the organizational structure puts up hurdles that make the overall equipment management process more effective.

Unsuitable for Changing Environment

The third problem is that most of the maintenance approaches are established based on a stable environment in which machines are installed and operated for a long period of time. In today's high-tech industries, especially in the semiconductor industry, in which chip capability grows exponentially and new products are introduced only months apart, the environment is dynamic, and equipment upgrades and new installations become almost routine.

In some cases, reliability theory and maintainability analysis cannot be applied because the equipment changes so fast that there is not enough data to come up with a meaningful statistical analysis. In other words, mean time between failures (MTBF) and mean time to repair (MTTR) do not have much meaning when the data sample size is not big enough. Similarly, expert systems do not have much practical value. Once the development is done and data builds up in the expert systems, the equipment may either be gone or have gone through a major upgrade.

Under the functional setup, any equipment change must be communicated to the maintenance and the operations groups by the engineering groups. The dynamic environment makes clear communication between these groups a huge challenge. Equipment changes also lead to skill upgrades for the personnel involved in managing the equipment. The functional approach prolongs the overall training process and the learning curve of these individuals.

In summary, the fundamental cause of the problems discussed in maintenance management lies in its traditions. For over a half century since the maintenance function was established, maintenance management has become a profession and a discipline. When a new factory is built, a maintenance department is formed without question. Although many improvements and developments have been made since the PM phase, they were within the box of the maintenance discipline. Out-of-the-box thinking is still absent. Generally, TPM is now being considered the ultimate solution for equipment management. What if further improvement is needed? How? The answer is breaking the maintenance traditions and seeking solutions beyond maintenance.

INTRODUCTION TO PHASE 6: POST-MAINTENANCE ERA

It is obvious that most of the existing approaches in the maintenance era cannot meet today's business requirements, especially in high-tech industries, in which business environments are dynamic. In the semiconductor industry, as demanded by the changes in operational scope and equipment characteristics, the traditional maintenance function had to be changed to push equipment performance to a higher level. The traditional incremental improvements made within the maintenance function could not meet the needs of the business environment. As a result, leading high-tech companies started at the turn of the century to challenge the old norms in equipment management. Fueled by the latest technologies and management theories, changes were implemented, which resulted in moving equipment management out of the maintenance era into the post-maintenance era [9,10].

From the organizational structure aspect, maintenance departments are disappearing, either absorbed by other departments such as operations and process engineering or replaced by new departments with titles such as equipment support engineering, equipment engineering services, and platform engineering. The word *maintenance* had disappeared from the daily operational vocabularies used in these leading manufacturing companies.

There is also a fundamental change in the responsibilities of such departments. Such responsibilities extend beyond traditional maintenance to include equipment development as well as utilization. These departments manage many aspects of the equipment process for which maintenance is only a small portion.

For instance, in the platform engineering setup, the employees of the departments are considered equipment platform managers and owners. In addition to keeping equipment up and running, they are responsible for equipment developments as well as implementations of equipment support structures, methodologies, and comprehensive computerized equipment management systems to provide a total solution in optimizing equipment output.

Before a machine enters the manufacturing facilities, these individuals have already worked with the equipment vendor to ensure that the performance specifications meet the manufacturing requirements. They conduct equipment acceptance at the vendor facilities as well as installation and acceptance in the final manufacturing plants. They also negotiate with the vendors on service contracts and support structures. They perform continuous improvement projects and equipment upgrades. They develop equipment training documents and train operations personnel. In addition, they are responsible for forecasting and controlling equipment budgets. They are also owners of automated and computerized equipment management systems. Equipment management becomes a streamlined process with a clear ownership. The consolidation of ownership also reduces miscommunication between the old functional groups and shortens the skill transfer process. These platform owners possess diverse technical skills as well as business administration skills. They are given the flexibility and power to ensure the best output regarding the equipment. These characteristics certainly usher the start of a new phase in equipment management.

At the same time, advanced IT continues to revolutionize the world. The single most important IT development in the 1990s was the Internet, which increased the speed of information exchanges. Historically, human society leaped to a new high when a new transportation system matured (i.e., prosperity followed the building of railroad system, highway system, and air travel system). Rather than transporting material and people, the Internet is a new type of transportation system that transports data and information at an unprecedented speed that continuously improves at a rapid rate.

The technology related to the Internet and intranet is an enabling tool that fuels the practices of new management concepts, such as reengineering and the circle of innovation, moving computerized equipment management systems to a new level of technical advancements. New generations of equipment management systems include advanced features such as multilevel performance tracking, real-time equipment status paging and messaging, automatic failure notification, and remote access to perform equipment repairs, all of which transform the field of equipment management into the post-maintenance era.

This book is the first attempt to define this new phase in equipment management called the post-maintenance era and to develop a framework

TABLE 3.2

An Extension of Equipment Management History

	Period	Characteristics	Objectives	Concepts Developed
Phase 6				
Post-maintenance era	2000s–	Total platform/ process focus Functional integration: vanishing maintenance department Automation and advanced computing/ communication applications	Optimization between availability and utilization Optimization of equipment development Users' satisfaction 100% maintenance free	Reengineering Circle of innovation Internet technology IM/text messaging Platform ownership Universal tech concept

that signifies it. The post-maintenance era serves as an extension of the pre-maintenance and maintenance era in equipment and is defined as phase 6. Part III of this book presents the post-maintenance era in great detail. The summary of this phase is shown in Table 3.2.

Section II

The Maintenance System

4

General Maintenance
Concepts and Practices

INTRODUCTION

Since the establishment of the field of maintenance in the 1950s, many concepts and practices have been developed and implemented, including preventive maintenance (PM), reliability-centered maintenance (RCM), predictive maintenance (PdM), maintenance prevention (MP), total productive maintenance (TPM), and terotechnology. These concepts and practices continue to be used in traditional industries as the mainstream principles in managing equipment. In high-tech industries, although companies are transitioning out of the maintenance era, many of these concepts and practices still serve as the basic building blocks of the new approaches.

PREVENTIVE MAINTENANCE

Preventive maintenance is the first major scientific approach to equipment management and started in the 1950s in the United States. PM has been defined as a series of tasks performed regularly either to extend the life of an asset or to detect the critical wear that causes an asset to fail [4]. It was often advocated as the sole means to reduce equipment failures and unplanned downtime.

PM was first started with mechanical equipment; the process is also called TLC: tighten, lube, and clean. As equipment becomes more complicated with computerization, common PM tasks include calibration, diagnostics, file backup, disk cleanup, defragmentation, and so on. The comprehensiveness of the PM tasks is one component of an effective PM

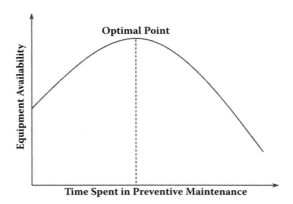

FIGURE 4.1
The relation between PM time and overall availability.

program. The other key components are the schedules for the PM and the resources needed for the PM. A PM schedule comprises how often and when PM should be performed. The resources for PM consist of PM staff, parts, tools, and documentation such as procedures and checklists.

Spending a great amount of time performing PM will certainly reduce unscheduled breakdowns. However, often PM means the equipment is not performing its intended function, which reduces equipment availability. Spending too little time performing PM will result in increased unplanned interruptions of equipment use. Unplanned breakdowns often result in longer downtime as resources such as staff and parts may not be readily available. Therefore, there is an optimal time spent in PM that results in the best equipment availability. Figure 4.1 demonstrates the relation between time spent in PM and overall equipment availability. To achieve the optimal time for PM, effective measures must be used in determining the PM schedule and then these must be continuously improved through regular reviews of the PM program as well as the amount of unscheduled breakdown events.

The following steps are recommended in setting up an effective PM program:

1. Gather PM information from the equipment manufacturer
2. Determine PM schedule
3. Determine PM task list
4. Determine PM staffing requirements
5. Determine PM equipment/tool requirements
6. Determine PM materials/parts requirements
7. Set up PM information system

The first step in developing a PM program is to obtain PM information from the equipment manufacturer. New equipment often comes with information on required PM to keep the equipment warranty valid or adhere to government regulations for safety reasons. The PM program developed and recommended by the equipment manufacturer may not be optimized for the user's operations since the usage models are different from function to function and even time to time. For instance, equipment may be purchased for engineering development operating Monday to Friday, 8 a.m. to 5 p.m., or for high-volume production running 24 hours a day and 7 days a week. Some equipment manufacturers offer different PM programs based on different usage models, so the user should work closely with the equipment manufacturer and negotiate the most suitable PM program for the equipment while keeping the warranty valid for the equipment. When the equipment is out of warranty and an extended service contract is not purchased, the user can then have more flexibility in determining the PM program as long as governmental regulatory requirements are met. The major components of a PM program are PM schedule and PM task list. Both can be optimally determined based on the equipment usage model and historic performance data.

Preventive Maintenance Schedule

The PM schedule consists of PM frequency and PM start time. PM frequency means how often PM should be performed. PM start time is the actual time PM is initiated. The most popular PM frequency method is time and calendar based. Typical equipment PM will be weekly, monthly, quarterly, semiannually, and annually. Some machines require daily inspection, which can be called a daily mini-PM. The calendar-based PM frequency is easiest to schedule and follow, but it does not take the effect of equipment usage into consideration. If a machine is idle most of the time, doing PM based on a clock may not be cost effective.

One alternative to calendar-based PM is PM based on output unit. An example of output unit-based PM is performing PM after every 10,000 units of product produced. Another example is automobile maintenance based on the mileage driven. This is the second most common method in PM frequency. The advantage of this method is that it is usage based, and usage is often directly correlated with wear and tear. It is also relatively easy to schedule if the production schedule is known, especially when production is operating in a predictable pattern. However, this method takes additional administrative effort and resources to monitor usage. It may also require additional hardware for measuring the output.

Output unit-based PM is suitable for a machine that takes one setup and a test run then starts to produce thousands of units, but in some cases, a machine may require frequent setups and verification runs to do small-size lots with different product mixes. Another alternative, called machine cycle-based PM, is suitable for these cases. In this method, the PM schedule is triggered when the equipment has gone through a certain number of operating cycles rather than producing a certain number of units. This method often requires meters to be installed on the equipment to monitor equipment cycle counts. As such, it allows PM to be done more precisely based on wear and tear as compared to the output unit-based method. On the other hand, it also takes more resources to set up the monitoring system or pay a premium for the equipment with the extra monitoring options. Also, it is more difficult to schedule the PM in advance.

An input unit-based method is another PM frequency-determining method that is based on equipment usage. In some equipment, the input is a more accurate measurement of machine usage, or it is easier to measure than the output and machine cycle counts. An example of this method is doing PM based on the utilities and energy used by the equipment, such as doing PM every time the gasoline tank requires refilling. Another example is performing PM based on the consumable depiction level, such as when the battery charge or the additive oil reaches a certain level. The input unit-based method has similar advantages and disadvantages as the machine cycle-based method except that it promotes utility usage consciousness for energy conservation. Also, depending on the equipment, using an input unit-based method may save costs and monitoring effort in yielding a more accurate measurement of machine usage.

Next, PM can also be scheduled based on output quality. The previously discussed output unit-based method focuses on the production output count, and the output quality-based method examines the yield or the trend of the production output quality. Typically, statistical control charts are used to monitor the defect rate of the products, and PM is triggered when the yield is lower than expected or when a worsening trend is apparent. This PM trigger method focuses on the bottom line of the operations, which ensures quality products exiting the factory and therefore contributes to customer confidence and satisfaction with the products. The disadvantage of this method is that it requires constant monitoring effort, and it is difficult to schedule in advance.

Last, PM can also be triggered by detecting certain equipment conditions. For instance, monitoring the vibration of a machine with mechanical parts can detect the critical point when the parts are beginning to fail. More details of this method are discussed in the section on PdM. This method aims to perform PM just in time (JIT) at the most optimal frequency without any waste. However, this method often requires significant effort in implementing monitoring systems and tracking the machine status in real time. Action is often required immediately once the worsening machine condition is detected. Therefore, it is similar to responding to unscheduled equipment failure, so it is difficult to schedule in advance. If resources are not ready to respond to this PM trigger, it may be too late, and the PM event becomes an equipment failure event. The different types of PM triggers are summarized in Table 4.1 with their advantages and disadvantages.

TABLE 4.1

Different Types of PM Triggers

Types of PM Frequency	Advantages	Disadvantages
Calendar based	Easy planning and scheduling PM resources	Does not account for equipment usage and product quality
Output unit based	Usage based and correlates to equipment wear; relatively easy to schedule with stable production schedule	Extra effort in monitoring and may require additional measurement hardware
Machine cycle based	Accurate machine usage based on machine operation	Difficult to schedule and requires extra systems for monitoring
Input unit based	Accurate PM scheduling on certain machines with less cost and monitoring effort; utility usage conscious for energy saving	Difficult to schedule and requires extra effort in monitoring
Output quality based	Quality driven and ensures quality products for better customer satisfaction	Very hard to schedule and requires constant effort in monitoring
Machine condition based	Catch failures right before they happen; just-in-time PM eliminated wastes for unnecessary scheduled downtime	Extremely hard to schedule and sometimes may even be too late as the condition already surfaced; also requires extensive monitoring systems

Once the PM frequency or trigger is determined, the next decision is when PM should be started. Often, equipment PM is scheduled around the activities of the operations. Weekends and off-hours are often considered because production or engineering activities are light. However, this may lead to longer downtime if unexpected parts are needed or vendor field service personnel are required since it usually takes a longer time to respond to part orders and field service calls during off-hours. Resources are another consideration when determining the PM start time. If vendor field service is required in certain PM activities, PM must be scheduled well ahead to accommodate the availability of the field engineers. This is often a dilemma with big PM events such as annual PM. An annual PM is often best scheduled during a factory shutdown, such as during holidays, but resources are often limited during those times as people are taking vacations. Also, higher cost is often required as extra pay is needed for people working on holidays.

Overall, an effective and efficient PM schedule must be determined based on the nature of the equipment, the operations, and the support structure, which includes the availability of parts and those providing labor. A regular review and modification of the PM schedule is also needed to ensure the PM schedule remains effective as equipment failure changes based on life cycle as well as usage model changes due to operational needs.

Preventive Maintenance Task List

Another main component of a PM program is the PM task list, which usually depends on the nature of the equipment. The most common PM tasks and examples are shown in Table 4.2. PM task lists come from a manufacturer's recommendations, government safety regulation requirements, industry standards, best-known methods, as well as internal requirements based on usage model. The determination of a PM task list should start with the equipment manufacturer's recommendations since the equipment manufacturer is the expert regarding its machines. The manufacturer's recommendations usually already include government regulatory requirements as well as industry standards. It is also typical that certain tasks must be performed to keep the equipment warranty valid. After taking the task lists from the manufacturer, one must review all regulatory requirements related to the equipment, especially in the context of state and local laws as most equipment manufacturers generate their requirements based on federal or international requirements. Next, industrial standards and

TABLE 4.2

Common PM Tasks and Examples

Common Tasks	Examples
Visual inspection	Check oil level
	Look for leaks
Data review	Check error log
	Interview operator/user
Diagnostics	Run self-check/diagnostic program
	Test a standard production unit
Cleaning	Remove dust
	Delete temp/old files from a disk
Reorganizing	Pull components back to default positions
	Defragment a hard disk
Tightening	Tighten bolts
	Tighten belts
Adjustment	Adjust pusher pressure
	Adjust sensor sensitivity
Lubrication	Lubricate motor bearings
	Lubricate joints
Replacement	Replace filter
	Replace seal
Calibration	Calibrate thermal sensor
	Calibrate voltage level and timing
Documentation	Record meter reading
	Issue PM compliance sticker

best-known methods should be considered. Joining equipment user groups and forums is recommended so up-to-date information can be obtained and fresh knowledge can be shared. Last, the equipment usage model and historical data must be considered. For instance, if a component has never failed in all weekly PMs, which means that the part has repeatedly been in good condition, the task of checking this component may not be needed in PM or may be considered for PM with a longer period. As such, equipment may have a PM task list that is the same with each PM trigger or may have different task lists for different PM, such as weekly, monthly, quarterly, semiannual, and annual PM. Unless the PM tasks are dictated by safety and warranty requirements, they should be based on an equipment usage model and performance data.

There are two approaches in developing PM procedures and PM checklists. One is a step-by-step detailed guide, and the other is a high-level task list that needs to be completed. The step-by-step detailed guide provides a complete description of the tasks and how the tasks are done. Drawings, illustrations, and decision flow charts are often seen in these procedures. They are usually found in highly structured factories, typically in a high-volume manufacturing environment with maintenance personnel with a lower skill level. Under such detailed PM procedures, tasks can be arranged serially, in parallel, or conditionally. Conditional statements such as "If oil level is above level 5, then skip steps 4 to 6 and go directly to step 7" are often used to guide inexperienced maintenance personnel. However, such an explicit list wastes time for experienced maintenance professionals, and giving them such a list to follow might be seen as an insult. Therefore, people who develop PM procedures and PM checklists should keep in mind the skill level of the maintenance personnel and provide suitable PM procedures to fit their skill levels. With skilled workers, the PM procedure can be merely a list of things to do without detailed sequential requirements and decision guidance.

Preventive Maintenance Staffing Requirements

A sufficient PM workforce with necessary skills is essential to the success of a PM program. Based on the PM task lists and the historical data on PM completion time, it is not difficult to identify the required skills and the workload. Therefore, the number of workers required to perform PM is easily obtained based on the workload and procedural requirements, such as a buddy system for heavy lifting or safety reasons.

Staffing with the right people for PM, however, takes more careful consideration. There are two approaches, and the first one is to put entry-level skill maintenance personnel on PM because PM tasks are typically rather routine and basic. Through the routine PM checkout, a technician with entry-level skill can learn the basics of the equipment, so it can be considered the starting point for the technician's career path in maintenance. The second method is having the best maintenance technicians perform PM as they are better at identifying potential issues to avoid equipment failures. Both approaches have valid reasons, and which one to use should depend on the nature of the equipment and the PM tasks. If the PM is simple and repetitive, workers with entry-level skills may be used. To double check if this is the right approach, the PM tasks must be able to be documented into a detailed step-by-step procedure and to be performed without much uncertainty and deviation. In some cases, the workers with entry-level skills can even be the operators of the equipment. On the other hand, if the equipment is complex, sensitive, and costly when it fails, it is best to staff the PM with the most skilled maintenance personnel.

Equipment skill level is not the only factor in selecting PM staff. Many maintenance professionals believe that the personal attributes for PM inspectors are different from repair technicians [4]. First, a good PM inspector is a diagnostician rather than a fixer. He or she must have a proactive mindset and must plan ahead. On the other hand, a repairperson is typically reactive and can do the job with minimal planning skills. Second, the PM inspector should have forecasting and statistical knowledge to be able to generate predictions and act on these predictions to prevent issues. Last, the PM inspector must be reliable, trustworthy, and able to work alone without close supervision since it is difficult to verify that some PM tasks actually were performed. In repairs, corrective actions are apparent when the equipment is back up to a running state. In PM, since the equipment is in working condition at the beginning, it is possible to take a shortcut in doing PM without anyone noticing. Hence, selecting the right people with the right personal attributes can increase the effectiveness of the PM program.

Another decision in staffing PM is whether to use a pool of maintenance personnel to conduct both PM and repairs or separate the maintenance personnel into a PM crew and a repair crew. In addition to the personal attribute differences regarding the PM inspector and repairperson mentioned, there is another reason to separate the team. Since equipment repairs are typically firefighting and have a higher priority than PM, having a separate PM crew can prevent interruptions of PM due to equipment repair calls. The disadvantage of having separate crews is the inefficiency in resource sharing. One crew may have nothing to do while another crew is very busy. Therefore, the maintenance manager should make this decision based on the size of the equipment base and the maintenance crew as well as the workload balance between PM and repairs.

Preventive Maintenance Equipment/Tool Requirements

PM tasks often require equipment and tools, from common mechanical hand tools, to sophisticated electronic scopes, to custom-made devices. The maintenance department must keep these equipment and tools in proper condition so that they can be readily used. Most organizations develop a standardized PM cart or a PM tool kit that consists of all the required equipment and tools for a specific PM so their maintenance personnel do not have to waste time gathering different tools. Also, many tools, such as meters and scopes, require regular calibration to ensure the accuracy of the measurements; keeping them calibrated is also essential. An oversight may cause an inability to do the PM, which leads to equipment failure or PM noncompliance.

Preventive Maintenance Materials/Parts Requirements

Similar to PM equipment and tools, required PM materials and parts must be readily accessible before PM starts. These include the consumable additives that need to be regularly filled, such as oils, nonreusable parts such as filters, and cleaning materials such as alcohol wipes. Developing a standardized PM parts kit is also highly recommended. The kit also serves as a confirmation check for completing all the PM tasks correctly as any leftover material in the kit should bring attention to the PM personnel that something must not be right. PM parts requirement management is part of maintenance spare and inventory management discussed in detail in Chapter 5.

Preventive Maintenance Information System

A PM information system includes all documentation related to the PM program, such as PM manuals, procedures, checklists, diagnostic logs, calibration files, passdowns, completion reports, certificates, stickers, and so on. When setting up the PM information system, management must keep the documentation process simple. Since PM is often a routine function, creating and keeping documentation become part of the routine tasks, and people seldom question the real value of the data. More often than not, the documents just pile up, and people rarely review them afterward. Therefore, a good PM record-keeping practice should start with reviewing the usefulness of each of the PM documents and then determine which documents are needed and for how long. Ideally, two categories of documents should be kept. The first type is documentations that are needed for compliance to safety regulations or ISO (International Organization for Standardization) 9000. The second type is the documentation that helps the organization to be more efficient in providing better maintenance of the equipment, such as a failure log or part replacement log. In general, management should attempt to combine these two types of documentations into one data depository to reduce duplication of administrative efforts in data entry and lookup.

The PM documents may be in electronic form or in physical hard copies. In today's electronic age, there is a tendency to move everything to electronic forms, but management must be careful that this does not create too many additional administrative tasks for the workers. Often, the equipment maintenance environment does not allow access to a computer when PM is performed on a machine. The maintenance technicians must use pen and paper, then do the data entry on a computer at a later time. Remember, a computer is simply a tool that is supposed to help an operation increase productivity. If using the tool becomes a burden to people, productivity will be impacted negatively.

Overall, PM is by far the most used mainstream maintenance tactic in equipment management. Setting up an effective PM program will benefit operations by reducing unscheduled failures and defects caused by the equipment failures. However, since PM is the oldest equipment maintenance concept and has been accepted as an absolutely must-have program for a long period of time, it has become rather routine and inherited much inefficiency over time. Further discussion of new approaches is presented in Part III of this book as part of the post-maintenance era.

RELIABILITY-CENTERED MAINTENANCE

Reliability-centered maintenance was developed by aircraft manufacturers, airlines, and government agencies in the productive maintenance phase when the reliability of the equipment became a key focus and safety was a primary consideration. It was launched in the U.S. commercial airline industry in the early 1960s and was first applied on a large scale for the maintenance of the Boeing 747 airplane [2]. RCM is a logical, structural, and systemic approach to equipment maintenance.

In general, RCM is a process that consists of identifying the equipment functions, causes of the functional failures, and the effects of the failures and then implementing the right tactics to prevent or cope with the failures. In practical implementation, different maintenance experts proposed several variations in the RCM process with various steps. For instance, Levitt defined that the RCM is a five-step process [4]. Moubray suggested that the RCM process entails asking seven questions about the asset [3]. Campbell described RCM as a recurring seven-step process that begins with an understanding of the business requirements and objectives [2]. Discussed next are the common steps given in most RCM literature.

The first step is to identify all functions of an asset [4]. It seems straightforward, but once the process is started, one often discovers that functions can have many facets. An asset may have primary, secondary, and protective functions, with each defined by a specific specification and performance standard. For instance, the pilot's seat in a fighter jet not only has the primary function of seating and holding the pilot with comfort but also has the secondary function of ejecting the pilot from the plane when the plane is damaged, plus providing a built-in parachute as well as a seat cushion for use as a floating device in the case of an emergency water landing. All functions must be considered, and user involvement is definitely essential in identifying all functions of an asset.

The second step is to find all ways the asset can lose its intended functions [4]. It is also called identification of functional failures or, in other words, the failure states [3]. A particular function can fail in numerous ways, typically shown as a total loss or partial loss of function. For instance, a moving mechanical component can lose its function if it stops moving completely, starts moving at a lower speed than specified, or begins moving in an undesired direction. Two key matters need to be addressed in this step; one is to have a clear boundary between acceptable and unacceptable performance, and the other is the understanding of the multiple levels of performance expected from the function. The boundary and level of performance are typically found in equipment specifications documents, but often the equipment specifications may not be the same as the actual requirements for processing the real products. While the equipment specifications provided by the equipment manufacturers are generic to all user conditions, a specific operation in a particular company generally has its own operation specifications, so both specifications must be considered.

The third step is to determine the failure modes and their causes and effects [2]. The failure modes are the events that are reasonably likely to cause the functional failures [3]. The list of failure modes typically includes failures caused by deterioration of the equipment and wear and tear. It should also include other failure modes, such as design flaws, operator errors, and maintenance oversights. Also, adequate analysis of the failure modes must be included to ensure that the causes of the failure modes are addressed rather than just the symptoms.

The fourth step is to identify and classify the consequences of each failure mode [4]. The consequences are often categorized and prioritized based on the degree of seriousness. The typical order of the consequences is loss of life, personal injury, environmental damage, and different levels of operational impacts, ranging from a line down in an entire factory, a significant amount of product damaged, and extensive equipment hard down, to lesser impacts such as short repairs and containments. Companies often rank the operational impact based on the costs of the consequences and the time needed to resolve the issues. Opportunity costs such as potential loss of revenue and market share are often part of the calculations rather than just the costs of fixing the problems. The consequences of the failures determine the intensity of the actions taken in the next step to prevent the failures or avoid the consequences.

The fifth step is to select feasible and effective measures to (1) detect the condition before failure to prevent it or (2) implement fail-safe backup systems to avoid the consequences when the failure occurs [2,4]. The measures include maintenance tactics such as PM and inspections, monitoring and alarm systems to signal vital conditions and failures, vigorous tests before release, and system redesign when the failures are still unacceptable after all preventive and corrective measures have been taken. When implementing such measures, management often faces budgetary decisions. It is costly to implement all possible measures to address the consequences of the failures. Successful implementation requires developing and identifying cost-effective options and selecting the options that (1) reduce the likelihood of multiple failures to an acceptable level, (2) are easier to implement, and (3) require less upkeep and recurring cost. For operational impact consequences, the cost of implementing the measures must be less than the production losses plus the repair costs.

While many experts conclude the RCM process at the fifth step, some include an extra step of continuous improvement (CI) and optimization of the implemented measures to refine the systems to be more effective in reducing failures and costs. Also, a few experts consider the RCM process as a recurring loop of steps with feedbacks. In any case, RCM takes extensive effort and commitment to implement. Not all areas and equipment have the need for RCM, so management should not roll out an RCM program without careful assessment of its operational requirements and constraints. An overview of the RCM implementation process is demonstrated in Figure 4.2.

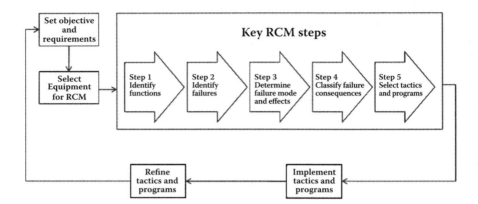

FIGURE 4.2
An overview of the RCM implementation process.

PREDICTIVE MAINTENANCE

Predictive maintenance is the application of predictive techniques to detect the condition of the equipment right before failure and, as a result, take measures to address the issue right before it occurs. Moubray listed over 50 PdM techniques, and new techniques are developed every year [3]. These techniques increase maintenance efficiency and save costs and time by doing the maintenance at the right time. For instance, a traditional PM program calls for changing an oil filter every quarter. Under PdM, the oil filter is changed when the differential pressure within the filter reaches certain readings. The technique used in this case is pressure monitoring. The popular categories of PdM techniques are chemical/particle analysis, vibration analysis, temperature monitoring, electrical monitoring, ultrasonic inspection, and advanced visual techniques.

Chemical and particle analysis is one of the most commonly used PdM techniques. The objective is to measure the changes in particle size and composition of the materials as a result of the deterioration of a system. According to Joel Levitt in his *Handbook of Maintenance Management*, there are seven basic types of chemical analysis: atomic emission spectrometry, atomic absorption spectrometry, gas chromatography, liquid chromatography, infrared spectrometry, fluorescence spectrometry, and thin-layer activation [4]. Some of these techniques measure the changing in particle size while others measure the foreign contents that signal wear or leakage.

Chemical and particle analysis is often used on lubricating systems, such as oil analysis, which is popular. In oil analysis, the reports typically show contamination materials such as dirt, sand, fuel, water, acids, bacteria, plastics, and metals in the units of parts per thousand (PPT), parts per million (PPM), or parts per billion (PPB). Diagnosis of system issues can be done based on the reported items. For example, a significant amount of metals in the oil probably indicates abnormal wear. An above-normal amount of fuel or water typically indicates that there is a leak in the system. Oil analysis often includes the study of the oil itself to determine if the oil has broken down or lost its viscosity over time. Often, the lubricating liquid is considered the lifeblood of a machine, so its chemical and particle analysis resembles blood tests for humans at a doctor's office. Other than liquid analysis, exhaust gas analysis is another method that can identify potential issues in certain equipment, kind of like a urine test for humans.

Vibration analysis is another widely used method in PdM in mechanical equipment and is typical in rotating machineries. Increasing vibration is often a signal for abnormal wear. Broadband analysis is the most popular method in vibration analysis [4]. It measures the changing amplitude of the vibration by frequency over time and plots it on an xy-axis chart. Large equipment like turbines and engines often has vibration transducers built in to feed the vibration measurements to the control systems. For most equipment without built-in vibration monitoring, the program is still relatively easy to set up if management is determined to apply this method. Portable vibration meters can be purchased or rented. Maintenance personnel can be trained to use the meters and set a regular schedule for the measurements. Management can then review the vibration readings to determine the maintenance actions based on the data.

Temperature monitoring is also commonly used because it can be applied to many types of machinery. It is typical for heat-related equipment such as furnaces, heat exchangers, boilers, steam turbines, and the like. It is also used in mechanical equipment as increasing friction leads to increasing temperature, which can indicate excessive wear, unbalanced loading, dull cutting tools, and so on. In electronic and electrical equipment, temperature monitoring can reveal poor contact or undesirable circuit resistance to identify failing components such as diodes, integrated circuit (IC) chips, transformers, switches, breakers, capacitors, batteries, and so on. Temperature-monitoring methods also range widely from infrared scanners, pyrometers, thermocouples, fiber-loop thermometry, still films, temperature transducers, heat-sensitive tapes, chalks, and more.

Electrical monitoring typically involves the measurement of the input amperage of the equipment. The increase in ampere input of the equipment often means that it requires more energy to overcome certain conditions, which may be abnormal. Constant reading of the input amperage can identify the behavior of the machine and correlate it with the machine operating modes and loads.

Ultrasonic inspection is a method that is increasingly applied in PdM. It is widely used in the medical field. Since there are many similarities between health care and equipment maintenance, it is inevitable that approaches and methods from the health care industry are absorbed for equipment maintenance use. Ultrasonic technology sends high-frequency sound waves into materials and picks up the echo to detect changes in the density of the materials. Therefore, it can determine the thickness of materials and identify corrosion, erosion, internal flaws, cracks, holes, pits, voids, and more. It is useful for monitoring enclosed areas in equipment where physical access is difficult, such as in pressurized pipes, filled tanks, and vacuum chambers. Another ultrasonic application in PdM is detecting ultrasonic waves generated by the mechanical systems as many times flows, leaks, bearing noise, loose fittings, and air infiltration produce highly directional ultrasonic waves [4]. Using the detectors, which translate high-frequency sound waves into hearable sound, can help to pinpoint the exact location of the issue.

Advanced visual techniques are also being used in PdM to identify issues to prevent failures. These techniques include the use of fiber-optic borescopes, ultrasmall video cameras, cold-light rigid probes, deep-probe endoscopes, pan-view fiber scopes, and so on [4]. These devices are often used for inspecting interior areas of equipment. Many devices are also equipped with built-in light sources and can be connected to a computer for imaging analysis that can flag small flaws, displacements, misalignments, foreign material buildup, and so on.

While many maintenance practices and approaches remain stable and unchanged for many years, PdM is actually one of the exciting areas in maintenance in which new techniques have been developed constantly as a result of the fast advancements in computing, vision, and software technologies for the past several decades. From a system point of view, any system has inputs of resources and outputs created by its internal processes. PdM ideas are simply generated by listing all the system inputs and its internal processes, and the outputs then find ways to monitor them. Analyzing the changes in these variables can lead to a increased understanding of the system's internal operational problems. Technologies already exist to enable equipment to give warning notifications and even take actions on their own based on the feedbacks from monitoring systems.

MAINTENANCE PREVENTION

Maintenance is often regarded as a series of activities that take place after equipment is made. The MP concept took maintenance activities upstream into the equipment design phase. *Maintenance prevention* is defined as a strategy to design and select machines with maintenance efficiency in mind to eliminate or reduce the need for maintenance. MP is also called early equipment management. It often involves incorporating data from current equipment improvements in new equipment design specifications.

The ideal objective in MP is to eliminate the need for maintenance completely by designing the equipment so robust that it will not fail during its life cycle. Examples can be seen in the automotive industry as some automobiles no longer require transmission services and timing belt changes for the life of the vehicles. The term *life cycle* is needed here since the machine designers must define the targeted life span of the machine to avoid overdesign. Today, many devices, especially in commercial electronics, are designed so that no maintenance is needed for the estimated period of usage, and if the devices do fail, it will be easier to discard them instead of repairing them. In addition to driving innovations into equipment functionalities and features, equipment designers are putting more focus on coming up with novel ideas to make equipment maintenance free.

While eliminating maintenance is the ideal state to aim for and may not be totally feasible in many cases, making maintenance easier becomes a common achievable target for equipment designers in most industries. The following tactics are often used in reducing maintenance efforts:

1. Design to reduce maintenance frequencies
2. Design to reduce maintenance steps
3. Design to provide easy access for maintenance
4. Design to reduce parts needed in maintenance
5. Design to use commonly available components
6. Design to standardize components

First, reducing maintenance frequencies means less scheduled downtime and lower upkeep cost. Many automobiles now have tune-up services extended to every 100,000 miles instead of every 30,000 miles as commonly seen in the 1970s. Many semiconductor machines no longer require weekly and even monthly PMs. Second, reducing maintenance steps means shorter maintenance time and lower labor costs. Most equipment has shorter PM lists that require less time to perform or have automation (using methods discussed in the PdM section) built in to enable self-monitoring and to reduce manual checkout and diagnosis.

Third, providing easy access for maintenance allows quicker and safer repairs and PMs. The placement of the components and the casing design of the equipment are carefully considered so each part is easy to access and replace. Proper clearance also is incorporated in equipment layout design. Fourth, reducing parts needed in maintenance leads to lower part and spare costs. For instance, some air filtering systems do not require filter changes.

Fifth, using commonly available components also enables quicker repairs and PMs by avoiding custom-made parts, which are typically more expensive and take a longer lead time to order. The examples are using off-the-shelf lightbulbs, cables, hoses, batteries, lubricants, and so on. Sixth, standardizing components can also save time in maintenance activities. An example would be using the same size of screws and nuts throughout all equipment so maintenance personnel do not have to switch tools in repairs and PMs, which also minimizes the need for purchasing multiple types of tools and stocking multiple types of similar parts. Dealing with multiple tools and multiple similar parts often adds confusion and extra handling steps in sorting out things, which leads to delays in repairs and PMs.

Applying these tactics mainly means taking maintenance considerations into the equipment design phase. One might ask how companies that have no ability to influence equipment manufacturers can practice the MP concept. MP also means making easy maintenance one of the key criteria in the equipment selection decision. The same objective is achieved by selecting equipment that is maintenance free. For many companies that already have the equipment in place, the concept can be carried out by redesigning or retrofitting the equipment to reduce maintenance requirements and maintaining an optimal operating environment for the equipment to minimize maintenance, such as keeping the area clean and removing sources of contamination.

TOTAL PRODUCTIVE MAINTENANCE

Total productive maintenance is a maintenance management approach that combines the American practice of PM with Japanese total quality management (TQM) and total employee involvement to achieve zero breakdown and zero defects. It can be viewed as an extension of TQM applied to the maintenance area. Many consider TPM a strategic philosophy rather than a program or an initiative. It is similar to the Toyota lean philosophy in eliminating wastes. The word *total* in TPM has three meanings: (1) It aims at total effectiveness: zero breakdown and defects; (2) it includes the practices of all previous maintenance concepts to maximize equipment efficiency and effectiveness; and (3) it emphasizes total involvement from all employees, from top management, equipment engineering, product engineering, operations, maintenance, purchasing, inventory management, facilities, IT/networking, to shipping and receiving [2].

Since Seiichi Nakajima published two books, *Introduction to TPM* and *TPM Development Program* in the 1980s [5,6], numerous Western writers have further elaborated on the concept and put it in the context of implementation in the Western world. According to John D. Campbell, TPM is based on three sets of principles: (1) maintenance engineering, (2) TQM, and (3) JIT [2]. Maintenance engineering includes the previously discussed maintenance concepts, such as PM, PdM, and MP, as well as equipment life-cycle costing and data management. The TQM concepts include employee participation and empowerment, CI, customer orientation, participative management style, and an open corporate culture. The JIT concepts focus on reducing cycle time, reducing lot sizes, eliminating non-value-added processes, and reducing all wastes (time, space, labor, inventory, materials, and movement), thus pushing companies to redesign operation processes, standardize procedures, and work closely with their suppliers.

Campbell continued to point out that TPM has three prime objectives: (1) maximizing equipment effectiveness and productivity through applying all maintenance concepts, (2) creating a sense of ownership in equipment operators through training and involvement programs, and (3) promoting CI through small-group activities involving all equipment-related personnel. To achieve the maximum effectiveness, which is zero breakdowns, five countermeasures are generally used [2]:

1. Maintain well-regulated baseline conditions.
2. Adhere to proper operating procedures.
3. Restore deterioration.
4. Improve design weaknesses.
5. Improve operation and maintenance skills.

Different enterprises have different visions and plans for TPM. Campbell provided a summary of seven elements and themes common in any TPM introduction: asset strategy, empowerment, resource planning and scheduling, system and procedures, measurement, CI teams, and processes. Asset strategy involves rearranging equipment and possibly eliminating redundant equipment to form more streamlined processes derived from JIT and TQM principles. Empowerment grants workers autonomy with new responsibilities. Operators are encouraged to participate in maintenance activities. Resource planning and scheduling must be carried out carefully as the introduction of TPM significantly increases demand on the maintenance personnel to train and produce procedures for the operators. System and procedures are required to transfer the maintenance knowledge and best practices effectively to the operators. A well-chosen data management system along with well-written procedures is instrumental for training and performing CI. Measurement refers to producing useful indicators for managing the progress of the CI efforts. CI teams provide effective forums to spark CI ideas and to promote a sense of ownership and leadership. As employees are empowered in generating CI ideas, the existing processes will certainly be revised, or new processes will need to be developed to ensure that TPM is carried out to its maximize benefits [2].

Wireman offered a slightly different but straightforward view of TPM. He pointed out that the focus of TPM is to become a low-cost producer by getting more out of the same assets than competitors. Furthermore, Wireman stated that the TPM philosophy is supported by four improvement activities: (1) improving maintenance efficiency and effectiveness; (2) focusing on early equipment management and MP; (3) improving the skills of all personnel involved with equipment; and (4) involving the operators in basic maintenance. Wireman also mentioned that the goal for TPM is to eliminate all equipment losses, which are breakdowns, setup and adjustment losses, idling and minor stoppage losses, startup and shutdown losses, reduced speed or capacity losses, and defects or reworks. Eliminating all these losses is beyond the ability of the maintenance department or any single department, which is why TPM emphasizes total involvement from all employees [11].

Joel Levitt, in his *Handbook of Maintenance Management* [4], highlighted that the key players in TPM are machine operators. Maintenance personnel take on the advisory role. Therefore, the key success factor to achieve TPM is to train the operators to reach fully autonomous maintenance. Based on the groundwork of Nakajima, Levitt summarized in seven steps the process that operators normally go through to achieve autonomous maintenance. These steps are carried out in small groups (called autonomous groups) settings. First, review the entire machine operation, learn and perform complete cleaning of the machine, tighten all fasteners to specifications, and repair any minor deficiencies that are apparent during the cleaning. Second, learn and perform inspections by following the manufacturer's manuals and studying equipment performance history. The group is also taught how to correct minor issues. Third, reduce the time to perform cleaning, remove the source of contaminations, and make the machine easier and quicker to service. Fourth, specify all tasks and frequencies and establish consistent standards for the tasks. The autonomous group prepares the documentation. Fifth, turn over inspection to the autonomous group. Maintenance personnel become coaches and are only involved in major problems. Sixth, systematize the autonomous maintenance activities and align the organization to support TPM. Seventh, achieve fully functioning TPM. Track results and efforts to provide ongoing recognition and feedback. Monitor failure frequency and look for CI opportunities [4].

TPM is typically implemented along with TQM and JIT as a package to change the culture and operational philosophy of the entire corporation [6]. Unlike other maintenance concepts, such as PM, RCM, and PdM, the maintenance department cannot act alone to implement the TPM concept as it calls for total involvement from all employees. Therefore, corporate executive management must set the direction to push for TPM. Also, since one of the key elements of TPM is CI, ongoing support from management is required to maintain the optimal success of TPM.

TEROTECHNOLOGY

Terotechnology was developed in the United Kingdom in the 1970s, about the same period of TPM development in Japan. It is defined as a combination of management, financial, engineering, and other practices applied to asset management to achieve economic life-cycle costs [2]. There are many similarities between terotechnology and TPM. Both have the objective of maximizing equipment effectiveness. Both are inclusive of different practices in management and engineering fields. Moreover, both demand involvement from other parties beyond maintenance personnel. The difference between terotechnology and TPM is that terotechnology is process oriented and emphasizes managing the entire equipment life cycle with processes of installation, commissioning, operation, maintenance, modification, and replacement. In practice, terotechnology is leaning toward involvement of the equipment supplier and engineering firms, whereas TPM puts most effort in user involvement.

5

Maintenance Management Logistics

INTRODUCTION

The key maintenance concepts described in Chapter 4 provide general directions to manage the maintenance operations. To manage the maintenance operations and apply the concepts effectively, maintenance managers need to master the tactical skills in planning and running the activities related to the logistics of the maintenance function. This chapter provides a practical guide to maintenance managers for dealing with the necessary logistics to manage the maintenance operations, such as planning and budgeting; developing the workers; dealing with customers, vendors, suppliers; and managing contracts and spare inventory.

PLANNING AND BUDGETING

Planning is one of the most important tasks for managers who run a functional area or department. Often, it is mandatory in well-structured corporations. Maintenance is no exception; in fact, since the maintenance department is considered an overhead function, its plan is typically under a great deal of scrutiny, with constant pressures for cost reduction. Therefore, maintenance managers often spend more effort in formulating and justifying the plan. A carefully developed and well-supported plan sets a good foundation for the success of the maintenance operation.

Planning includes strategic and tactical planning. Strategic planning includes the direction of the department along with headcount and budget requirements. Most corporations have annual planning cycles set up when all departments need to submit their headcount and budget plans for the next year. Managers are also required to participate in quarterly planning update cycles to update the progress and explain any deviations from the original annual plan. Tactical planning is related to task scheduling, excursion handling, labor assignments, and materials arrangements.

Strategic Planning in Maintenance

Strategic planning in maintenance is typically demonstrated in the long-range strategic plan and the annual plan for the maintenance department. The long-range strategic plan typically aims to provide a business outlook of 3 to 5 years. Since maintenance is a department with rather stable overhead in many companies, maintenance management is seldom required to provide a long-range strategic plan. The major strategic planning activity for maintenance concentrates on the annual planning and its update cycles. The annual plan for the maintenance department should include the following sections:

1. Summary (missions, goals, changes, assumptions)
2. Key challenges and opportunities
3. Efficiency improvements
4. Risks and mitigation tactics
5. Headcount requirement
6. Budget requirement

The summary section should include mission, objectives and goals, changes, and assumptions. First, the mission of the operation should be stated. The mission statement sets the direction of the department. Second, the objectives and goals for the planning cycle should be presented. Objectives and goals are different from mission as they need to be measurable and often more than one statement. They are the building blocks for achieving the mission. The mission statement typically does not change every planning cycle, but the goals and objectives should be specific to what the department needs to complete in that particular planning cycle. Third, any changes that have an impact on the next planning cycle should be highlighted. Examples of the changes are the increase or decrease of equipment count, introduction of new equipment, implementation of new maintenance concepts, major restoration or upgrade to equipment, change in support personnel, change in vendors and suppliers, and change in warranty and contracts. Last, any assumptions that serve as the basis of the plan should be stated. The maintenance business is associated with high uncertainty as unscheduled equipment failures occur by chance. The current year may be a good year with minimal downtime and low maintenance spending, but it does not mean next year will be the same. Therefore, assumptions, such as stable reliability and maintainability indicators, often need to derive the plan. Changing assumptions can significantly change the planned headcount and budgets, so all the assumptions need to be made clear to upper management before presenting the numbers.

The key challenges and opportunities section should first point out the difficulty factors in performing maintenance in the next planning cycles and then present the opportunities to improve the operations. A challenge often brings opportunities to take the organization to the next level of improvement, so these go together. For example, an organization may face the challenge of keeping up an end-of-life machine with frequent breakdowns. It could be an excellent opportunity to retool or perform a major overhaul of the equipment. Another example is that the maintenance department may be challenged to maintain the same level of resources but to support an increasing equipment base. This may be a great opportunity to implement total productive maintenance (TPM) and transfer simple maintenance tasks to the operations department.

The business environment forces companies to improve their efficiency year after year to stay competitive. Maintenance as an overhead function is often the focus and the starting point for cost cutting. Maintenance managers should include a section in the plan outlining all the measures in the next planning cycle to increase equipment uptime, reliability, and maintainability as well as reduce maintenance costs. Examples of efficiency improvement programs are upgrading skill sets of maintenance and operations personnel, developing new repair and preventive maintenance (PM) procedures, negotiating new contracts with new vendors, changing spare management strategy, and the like. This section demonstrates proactiveness and avoids the perception that maintenance is reactively waiting for equipment failures to happen and to react to them.

Since maintenance deals with statistical uncertainties, the risk factor is higher than for other operations. The plan should identify the proper risks and generate corresponding mitigation tactics. The risk items are typical but not limited to the following: significant equipment excursions; potential safety and ergonomic issues; process changes that demand new equipment capacity and capability; vendor support changes (such as termination of the warranty and contract, business closure, change of ownership, reduction of field service as a result of layoff, etc.); spare and materials availability and pricing inflations; new software environment (such as patches and upgrades required to keep the equipment in compliance with the latest standards or network connectivity and security protocols); as well as internal equipment support structure changes (such as loss of key personnel and implementation of a new computerized maintenance management system, CMMS). The objective of putting a risk management section in the plan is to anticipate issues to avoid firefighting and unnecessary stresses that may cause the operation to be out of control.

The headcount and budget requirements are the main essences of the plan. Numbers for both must be derived from solid data to be justifiable. The maintenance business is never steady, so defining the proper support level is key. The decision should be based on historical downtime data, vendor recommendations, production requirements, as well as operations inputs. The headcount plan should be done first as the labor costs in the budget plan are based on the headcount.

Headcount Plan

Headcount is based on the workload. The main work of the maintenance group is scheduled and unscheduled maintenance work. Typical scheduled work is PM, upgrades, and setups. Unscheduled work is repairs, assists, and some customer services. Figure 5.1 shows a spreadsheet example of how headcount can be calculated based on the equipment information. The data cells with a dotted pattern and italic numbers are the user input information. The following are explanations to guide the data entry for these cells:

- Cell B1: The type of the equipment. Similar equipment with similar maintenance historical data can be grouped together.
- Cell F1: The year of the planning cycle.
- Cells C3–F3: Number of the equipment by quarter.
- Cell B4: The 13-week mean time between failures (MTBF) in hours. It is typical to use the 13-week MTBF instead of a 4-week or weekly MTBF because it fluctuates less and is good for annual planning. If historical data are not available, ask the equipment manufacturer for a recommendation. The same applies for all the "mean time" numbers discussed.

	A	B	C	D	E	F
1	Equipment:	*Example System*			Year:	*2010*
2		Annual Avg.	Q1	Q2	Q3	Q4
3	Equipment count:	9	8	8	10	10
4	13-week MTBF (hour):	500		13-week MTBF (hour):		96
5	13-week MTTR (hour):	3.5		13-week MTTR (hour):		1
6						
7	Activities	Hour/each	Frequency	Hour/year	% Support	Actual Time
8	PM					
9	Annual	10	1	10	2%	0.2
10	Semiannual	8	1	8	2%	0.2
11	Quarterly	6	2	12	50%	6
12	Monthly	4	8	32	90%	28.8
13	Weekly	2	40	80	100%	80
14	Daily	0.5	313	156.5	100%	156.5
15				Total Annual PM Time:		271.7
16	Setup					
17	Major Setup	3	12	36	100%	36
18	Minor Setup	0.5	104	52	50%	26
19				Total Annual Setup Time:		62
20				Total Scheduled Time:		333.7
21	Repairs	3.5	17.5	61.3	90%	55.2
22	Assists	1	91.3	91.3	50%	45.6
23				Total Unscheduled Time:		100.8
24				Total Maintenance Work per System:		434.5
25						
26	Headcount Requirement	% of HC	Q1	Q2	Q3	Q4
27	Maintenance Work	60%	869	869	1086	1086
28	CI Projects	5%	72	72	91	91
29	Non-linear Workload	5%	72	72	91	91
30	Training	5%	72	72	91	91
31	Meeting/Passdown	5%	72	72	91	91
32	Spares Handling	2%	29	29	36	36
33	Breaks	6%	87	87	109	109
34	Vacation/Personal Absence	12%	174	174	217	217
35	Total Activity Hours	100%	1448	1448	1810	1810
36	Total Headcount Requirements:		2.8	2.8	3.5	3.5

FIGURE 5.1
Headcount calculation worksheet example.

- Cell B5: The 13-week mean time to repairs (MTTR) in hours.
- Cell F4: The 13-week mean time between assists (MTBA) in hours. Not all equipment issues are related to failures (i.e., an equipment lockup and cleared without finding a problem). Assists are more often the failures.

- Cell F5: The 13-week mean time to assists (MTTA) in hours.
- Cells B9–B14: Average hours for each type of PM per equipment. Information can be obtained from the equipment manufacturer if historical data do not exist.
- Cells E9–E14: Percentages of time maintenance personnel are involved in supporting the PM. In the example, it assumes that annual and semiannual PM is done by the vendors and only minimal administrative time is involved (such as calling the vendors and accompanying them to the equipment location, etc.) As the frequency of PM increases, the internal maintenance personnel take on more ownership. Weekly and daily PM is completely owned by the internal maintenance department.
- Cells B17–B18: Average hours for setups per equipment. If historical data do not exist, obtain input from product engineering, operations, maintenance techs, as well as the equipment manufacturer.
- Cells C17–C18: Estimated number of times setups need to be done in a year.
- Cells E17–E18: Percentages of time maintenance personnel are involved in supporting the setups. The example assumes that maintenance personnel perform all major setups and share half of the minor setups with operators and users.
- Cells E21–E22: Percentages of time maintenance personnel are involved in supporting the unscheduled repairs and assists. The example assumes that maintenance personnel perform most repairs, with a few difficult ones done by vendors, and half of the assists are shared with the operations department.
- Cells B27–B34: Percentages of time for all maintenance activities, including overhead and administrative tasks. Changing the percentage allocation between the activities will change the headcount requirements.

Further explanation is needed regarding the percentage allocation of the maintenance activities. Depending on the nature of the business as well as company customs and operations, the maintenance work typically counts for 50% to 70% of the total work activities. Continuous improvement (CI) projects in equipment are common but may be not applied to all companies.

Nonlinear workload is used to create a buffer for the operations that have large workload variations. Without the consideration of the nonlinear workload, the headcount will not cover some busy days. In theory, since averages are being used in the calculations, half of the days have a below-average typical workload, and half of the days have an above-average workload. Although some administrative activities, such as training, meetings, and vacations, can be loaded in the days that have a below-average workload, not all tasks can be scheduled as maintenance intrinsically has many uncertainties; in most cases, the over-workload days are not predictable. Therefore, there will be many days in a year when the resources are not enough to cover the equipment activities. To reduce the numbers of these "crazy" days and minimize the stress to the department, use the nonlinear workload allocation to create a buffer. Based on the historical data and forecast of the equipment issues in the planning cycle, managers will become increasingly more adept at allocating the right percentage over time.

Training includes formal equipment training by the vendors and the peer-to-peer on-the-job training. It also includes non-equipment-related training, such as corporate required trainings that apply to all employees. Meetings and passdown are used for corporate meetings, information exchange meetings between shifts and departments, as well as the time for documenting equipment information to pass down the data to peers, users, and managers. Spare handling is the time needed for placing orders, receiving, incoming inspecting, stocking new parts, tagging and sending failed parts for repairs, checking for returning status, and so on.

Breaks are accounted at 6% in the example, which is about a 15-minute break for each 4 hours of work. Lunch breaks are not counted as typically they are unpaid time for the employees. In some cases if particular operations have a long downtime and PM time and maintenance techs are not covering for each other while on break, the downtime hours may already be included in some of the break time. The percentage needs to be adjusted down. Vacation and personal absence time are normal to all businesses, and managers should have the data on the average time that the employees are out of the plant in a year. The 12% used in the example is typical and equates to about 6 weeks of vacation, sickness, and personal time off in a year. Different operations may have other categories that are not shown in this example. For example, some companies require maintenance personnel to participate in an emergency response team (ERT) and business recovery planning. Adequate headcount allocations are needed to accommodate such activities.

Now, let us examine the cells with formulas in the spreadsheet:

- Cell B3: Annual average count for the given type of the equipment. If the headcount plan of the company does not require quarterly breakouts, this number can be used for the headcount requirement for the entire year. The formula is B3 = AVERAGE (C3:F3).
- Cells C9–C14: Number of times each PM is performed in a calendar year. The example assumes skipping a higher-frequency PM when a lower-frequency PM is due; that is when a monthly PM is due, the weekly PM and daily PM will not be performed on that day.
- Cell B21: Average repair time; same as MTTR in Cell B5.
- Cell C21: Estimated number of times repairs are performed in a calendar year. Calculated based on MTBF. The formula is C21 = 24*365/ B4. For low MTBF and high MTTR numbers, the total repair hours need to be deduced from the total annual hours (24*365) to produce a more accurate estimation.
- Cell B22: Average repair time; same as MTTA in Cell F5.
- Cell C22: Estimated number of times assists are performed in a calendar year. Calculated based on MTBA. The formula is C22 = 24*365/F4. Same as cell C21; deduct the total annual assist hours from the total annual hours if the MTBA is low and MTTA is high.
- Cells D9–D14, D17–D18, and D21–D22: Annual hours required in a calendar year. The formula is D# = B#*C#, where # is the corresponding row number.
- Cells F9–F14, F17–F18, and F21–F22: Actual annual maintenance support hours required in a calendar year. The formula is F# = D#*E#, where # is the corresponding row number.
- Cell F15: Total annual PM time by maintenance personnel. The formula is F15 = SUM (F9:F14).
- Cell F19: Total annual setup time by maintenance personnel. The formula is F19 = SUM (F17:F18).
- Cell F20: Total annual scheduled time by maintenance personnel. The formula is F20 = F15 + F19. If other scheduled events such as upgrades are done by the maintenance department, add rows to include them in the total.
- Cell F23: Total annual unscheduled time by maintenance personnel. The formula is F23 = SUM (F21:F22).
- Cell F24: Total annual time by maintenance personnel. The formula is F24 = F20 + F23.
- Cells C27–F27: Maintenance work hours by quarter based on the number of equipment per quarter shown in row 3. The formula is X27 = $F24/4*X3, where X is the corresponding column letter.

- Cells C28–F34: Total hours maintenance department spends on each activity per quarter. The formula is X# = \$B#/\$B\$27*X\$27, where X is the corresponding column letter and # is the corresponding row number.
- Cells C35–F35: Total quarterly hours required by the maintenance department for all activities. The formula is X35 = SUM (X27:X34), where X is the corresponding column letter.
- Cells C36–F36: Estimated maintenance headcount required for the given type of equipment per quarter. The example assumes each worker works 40 hours a week. The formula is X36 = X35/(40*13), where X is the corresponding column letter.

The example spreadsheet covers only one type of equipment. A factory typically has many different types of equipment, so one sheet for each type of equipment should be used. The total headcount for the department should be the sum of all spreadsheets. Management overhead, such as managers and administrative assistants, should also be added to the total headcount. If the operations are running 7 days a week and 24 hours a day with different shifts, the shift supervisors should be added as well.

Budget Plan

Equipment budgets are typically separated into capital and expense due to tax laws. Equipment purchases, large upgrades, and equipment installation costs are typically capital as companies can claim depreciation and obtain tax benefits. Different industries and different companies have their own policies that comply with tax laws and dictate what costs should be capital. Maintenance managers need to work with the company's finance department to ensure the budget is done correctly. Most of the costs associated with maintenance are expenses. Figure 5.2 demonstrates an example of an expense budget worksheet for a particular type of equipment. Each equipment type should have an individual worksheet, and all the worksheets should be rolled up to the department-level budget, which is demonstrated in Figure 5.3. Also, please note that expense cost entries are separated into two columns, fixed cost and variable cost. As mentioned, the maintenance operations are inherently associated with high uncertainty. Putting everything in a single bucket will lead to a large gap between forecast budget and actual spending, which is not what the finance department and upper management want to see. It gives a perception that the department

Example System

ITEMS	Q1 PLAN			Q2 PLAN			Q3 PLAN			Q3 PLAN			2010 Total		
	Fixed	Var.	Total	Fixed	Var.	Total	Fixed	Var.	Total	Fixed	Var.	Total	Fixed	Var.	Total
Equipment Training															
Vendor training	$2,000		$2,000			$0	$2,000		$2,000			$0	$4,000	$0	$4,000
Travel and logging	$2,000		$2,000			$0	$2,000		$2,000			$0	$4,000	$0	$4,000
In-house training			$0	$200		$200			$0	$200		$200	$400	$0	$400
Training Total	$4,000	$0	$4,000	$200	$0	$200	$4,000	$0	$4,000	$200	$0	$200	$8,400	$0	$8,400
Equipment Spare Parts															
Purchased spares	$5,000		$5,000		$3,000	$3,000	$10,000		$10,000		$3,000	$3,000	$15,000	$6,000	$21,000
Consumable parts	$800	$200	$1,000	$800	$200	$1,000	$1,000	$300	$1,300	$1,000	$300	$1,300	$3,600	$1,000	$4,600
Consignment fees	$1,000		$1,000	$1,000		$1,000	$1,200		$1,200	$1,200		$1,200	$4,400		$4,400
Spare maintenance costs		$5,000	$5,000		$5,000	$5,000		$6,000	$6,000		$6,000	$6,000	$0	$22,000	$22,000
Spare management fees	$5,000	$0	$5,000	$5,000	$0	$5,000	$5,000	$0	$5,000	$5,000	$0	$5,000	$20,000	$0	$20,000
Spare Total	$11,800	$5,200	$17,000	$6,800	$8,200	$15,000	$17,200	$6,300	$23,500	$7,200	$9,300	$16,500	$43,000	$29,000	$72,000
Equipment Contract															
System#1 SN:10001	$4,500		$4,500	$3,500		$4,500	$4,500		$4,500	$4,500		$4,500	$18,000		$18,000
System#2 SN:10005	$3,000		$3,000	$4,500		$4,500	$4,500		$4,500	$4,500		$4,500	$18,500		$18,500
System#3 SN:10012			$0	$3,000		$3,000	$4,500		$4,500	$4,500		$4,500	$12,000		$12,000
System#4 SN:10014			$0	$3,000		$3,000	$4,500		$4,500	$4,500		$4,500	$12,000		$12,000
System#5 SN:10025			$0	$1,500		$1,500	$4,500		$4,500	$4,500		$4,500	$10,500		$10,500
System#6 SN:10025			$0	$1,500		$1,500	$4,500		$4,500	$4,500		$4,500	$10,500		$10,500
System#7 SN:10031			$0			$0			$0			$0	$0		$0
System#8 SN:10033			$0			$0			$0			$0	$0		$0
System#9 SN:TBD			$0			$0			$0			$0	$0		$0
System#10 SN:TBD	$0	$0	$0			$0			$0			$0	$0		$0
On-site field service	$30,000		$30,000	$30,000		$30,000	$30,000		$30,000	$30,000		$30,000	$120,000		$120,000
24x7 On-call support	$3,000		$3,000	$3,000		$3,000	$3,000		$3,000	$3,000		$3,000	$12,000		$12,000
Billable	$0	$5,000	$5,000		$5,000	$5,000		$5,000	$5,000		$5,000	$5,000	$0	$20,000	$20,000

Contract Total	$40,500	$5,000	$45,000	$51,000	$5,000	$56,000	$60,000	$5,000	$65,000	$60,000	$5,000	$65,000	$211,000	$20,000	$231,000
Support Tools															
Tool purchasing	$5,000		$5,000						$0		$2,000	$2,000	$5,000	$2,000	$7,000
Tool rental	$1,000		$1,000	$1,000		$1,000	$1,000		$1,000	$1,000		$1,000	$4,000	$0	$4,000
Tool Total	$6,000	$0	$6,000	$1,000	$0	$1,000	$1,000	$0	$1,000	$1,000	$2,000	$3,000	$9,000	$2,000	$11,000
Software and Licenses															
Support software	$1,000		$1,000	$1,000		$1,000	$1,000		$1,000	$1,000		$1,000	$4,000	$0	$4,000
License Fee			$0			$0	$2,500		$0	$2,500		$2,500	$2,500	$0	$2,500
Software and Licenses Total	$1,000	$0	$1,000	$1,000	$0	$1,000	$1,000	$0	$1,000	$3,500	$0	$3,500	$5,500	$0	$5,500
Upgrade/CI Project															
Upgrade materials			$0	$5,000	$2,000	$7,000			$0			$0	$5,000	$2,000	$7,000
Upgrade labor			$0	$2,000	$1,000	$3,000			$0			$0	$2,000	$1,000	$3,000
Upgrade/CI Project Total	$0	$0	$0	$7,000	$3,000	$10,000	$0	$0	$0	$0	$0	$0	$7,000	$3,000	$10,000
Facilities charges															
Facilities usage	$800	$200	$1,000	$800	$200	$1,000	$1,000	$200	$1,200	$1,000	$200	$1,200	$3,600	$800	$4,400
Facilities modifications	$500	$200	$700	$500	$200	$700	$500	$200	$700	$500	$200	$700	$2,000	$800	$2,800
De-installation/move charges			$0			$0			$0			$0	$0	$0	$0
Facilities Charges Total	$1,300	$400	$1,700	$1,300	$400	$1,700	$1,500	$400	$1,900	$1,500	$40	$1,900	$5,600	$1,500	$7,200
Total	$64,600	$10,600	$75,200	$68,300	$15,600	$84,900	$84,700	$11,700	$96,400	$73,400	$16,700	$90,100	$291,500	$55,500	$346,500

FIGURE 5.2

An equipment expense budget worksheet example.

($ in thousands) ITEMS	Q1 PLAN Fixed	Var.	Total	Q2 PLAN Fixed	Var.	Total	Q3 PLAN Fixed	Var.	Total	Q3 PLAN Fixed	Var.	Total	2010 Total Fixed	Var.	Total
Payroll															
Base pay	1,200		1,200	1,200		1,200	1,320		1,320	1,320		1,320	5,040	0	5,040
Overtime pay		24	24		48	48		26	26		26	26	0	125	125
Benefits	60		60	60		60	66		66	66		66	252	0	252
Bonus		60	60		60	60		66	66		66	66	0	252	252
Payroll Total	1,260	84	1,344	1,260	108	1,368	1,386	92	1,478	1,386	92	1,478	5,792	377	5,669
Supplemental Resources															
Consulting	2	1	3		2	2	2	1	3		2	2	4	6	10
Temporary work		20	20		20	20			0			0	0	40	40
Contracting work	40		40	40		40	40		40	40		40	160	0	160
Supplemental Resources Total	42	21	63	40	22	62	42	1	43	40	2	42	164	46	210
General spending															
Safety supplies	2		2	2		2	2		2	2		2	8	0	8
Communication	2		2	2		2	2		2	2		2	8	0	8
Team events	1		1	1	0.5	1.5	1		1	1	2	3	4	2.5	6.5
Miscellaneous supplies	2	1	3	3	1	3	2	1	3	2	1	3	8	4	12
General Spending Total	7	1	8	7	1.5	8.5	7	1	8	7	3	10	28	6.5	34.5
Training															
General training	2		2	2		2	2		2	2		2	8	0	8
Equipment1	4	0	4	0.2	0	0.2	4	0	4	0.2	0	0.2	8.4	0	8.4
Equipment2			0			0	2		2			0	2	0	2
			0			0			0	2		0	0	0	0
Training Total	6	0	6	22	0	27	8	0	8	2.2	0	22	18.4	0	13.4
Equipment Spares/Parts															
General spares purchasing	2	1	3	2	1	3	2	1	3	2	1	3	8	4	12
Equipment1	11.8	5.2	17	6.8	8.2	15	17.2	6.3.2	23.5	7.2	9.3	16.5	43	29	72
Equipment2	15	10	25	15	10	25	15	10	25	15	10	25	60	40	100
			0			0			0			0	0	0	0
Spare Total	28.8	16.2	45	23.8	19.7	43	34. 7	17.3	51.5	24.2	20.3	44.5	111	73	184
Equipment Contract															
Equipment1	40.5	5	45.5	51	5	56	60	5	65	60	5	65	211.5	20	231.5
Equipment2		15	15		15	15		15	15		15	15	0	60	60
			0			0			0			0	0	0	0
Contract Total	40.5	20	60.5	51	20	71	60	20	80	60	20	80	211.5	80	291.5
Support Tools															
General tool purchasing	10	1	11	5	1	6	10	1	11	5	1	6	30	4	34
Equipment1	6	0	6	1	0	1	1	0	1	1	2	3	9	2	11
Equipment2		0.2	0.2		0.2	0.2		0.2	0.2		0.2	0.2	0	0.8	0.8
			0			0			0			0	0	0	0
Tool Total	16	1.2	17.2	6	1.2	7.2	11	1.2	12.2	6	3.2	9.2	39	6.8	45.8
Software and Licenses															
General software and licenses	3	1	4	3	1	4	3	1	4	3	1	4	12	4	16
Equipment1	1	0	1	1	0	1	1	0	1	3.5	0	3.5	6.5	0	6.5
Equipment2			0			0			0	3		3	3	0	3
			0			0			0			0	0	0	0
Software and Licenses Total	4	1	5	4	1	5	4	1	5	9.5	1	10.5	21.5	4	25.5
Upgrade/CI Project															
General CI projects	10	1	11	5	1	16	10	1	11	5	1	6	30	4	34
Equipment1	0	0	0	7	3	10	0	0	0	0	0	0	7	3	10
Equipment2			0			0			0			0	0	0	0
			0			0			0			0	0	0	0
Upgrade/CI Project Total	10	1	11	12	4	16	10	1	11	5	1	6	37	7	44
Facilities Charges															
General facilities charges	5	1	6	5	1	6	5	1	6	5	1	6	20	4	24
Equipment1	1.3	0.4	1.7	1.3	0.4	1.7	1.5	0.4	1.9	1.5	0.4	1.9	5.6	1.6	7.2
Equipment2	0.2		0.2	0.2		0.2	0.2		0.2	50	5	55	50.6	5	55.6
			0			0			0			0	0	0	0
Facilities Charges Total	6.5	1.4	7.9	6.5	1.4	7.9	6.7	1.4	8.1	56.5	6.4	62.9	76.2	10.6	86.8
Total	1,421	147	1,568	1,413	178	1,591	1,569	136	1,705	1,596	149	1,746	5,999	611	6,609

FIGURE 5.3

A department budget summary worksheet example.

spending is out of control. By separating the budget into fixed and variable costs, it demonstrates that fixed costs can be tracked closely to the plan target; therefore, the department is financially well managed apart from the uncertainties in its operating environment. Variable costs should still be forecast based on historical data and attempting to hit the target as close to the plan as possible.

For each individual type of equipment, the expense budget for maintenance typically consists of training, spares and parts, contracts, support tools, software and licenses, upgrades and CI projects, and facilities charges. Training cost includes money paid for vendor training, travel and lodging, and in-house training. Vendor training cost is paid to the equipment vendors or outside training firms. Sometimes, training credits are accompanied with new equipment purchases, so please check before making an entry. Travel and lodging expenses are for people attending vendor classes or classes to be held off site and in other factory locations. In-house training cost typically covers training materials and setup expenses.

Spare and part cost covers the purchase of spares, consumable parts, consignment fees, spare repairs and maintenance cost, as well as any spare management fees. Purchasing spares can reduce equipment downtime as failed parts can be replaced quickly without waiting for order and delivery. It is a big part of the maintenance expense budget. Consumable parts are the nonrepairable parts that are required for PM and regular refill and replacement, such as filters, lightbulbs, gaskets, coolants, and the like. Since purchased spares are costly, some companies are negotiating with the vendors to stock spares locally for a fee, which is called a consignment fee. It is for renting spares essentially. Typically, it is 2–5% of the value of the spares on a monthly basis. Spare repair and maintenance cost is the cost to pay for repairing the spares or keeping the spares in a ready condition. Spare management fees are the money paid to audit and stock the spare inventory as some companies outsource their spare management function.

The equipment contract cost section covers the cost for extended warranties or time and materials charges. Extended warranties should be under fixed cost, and time and materials should be under variable cost. To produce a more accurate budget plan, the charges should be broken down to each piece of equipment as different configurations may have different warranty prices even if they belong to the same type of equipment. Also, since the timings of equipment installed in the factory are generally different, the machines have different out-of-warranty dates, so the cost might be different. This section also includes cost for on-site field service, such as payment for putting a vendor field service engineer (FSE) to work on site. Next, managers need to consider the cost that the vendor charges to put an around-the-clock, 7-day-a-week on-call support infrastructure together and the billable charges for the on-call service when the FSE comes on-site during off hours and weekends. The charges are typically by hour with a 4-hour minimum.

In the supporting tool category, the costs for purchasing tools required to perform PMs or repairs needs to be included. Some expensive tools, such as digital scopes and predictive maintenance monitoring tools, may be considered capital equipment, so they should be in the capital budget. Managers need to check with the finance department for guidance. The support tool costs also include tool rental cost. In some cases, renting may be a better choice when the tools are too expensive and the usage period is short.

Some machines require software and licenses to perform their functions. Software costs include the money spent to purchase diagnostic software, calibration software, networking and automation software, and predictive maintenance monitoring and analyzing software. Licensing fees are sometimes required for using the software, so they need to be included in the maintenance budget.

The upgrade and CI project cost category is for material and labor charges to perform equipment improvement projects. These projects are typically related to improving the performance, safety, ergonomics, and maintainability of the equipment.

The last cost category is associated with facility charges. It includes the costs for facilities such as chemicals, gases, and liquid nitrogen needed to calibrate or repair the equipment. Usually, the facility usage charges do not include space rental, electricity, and water as these are commonly included in the company's general budget and not in the departmental budget. Facility modification charges are related to the cost of changing equipment facility setups, such as switching and changing fittings, chemical supply tanks, and gas cylinders. When the equipment is at the end of life or needs to be moved to a different location, deinstallation and moving costs are also part of the facility expense category. Equipment installation costs are typically capital spending, but deinstallation and moving costs are expense spending. These charges include labor costs for facilities disconnect, equipment cleaning, chemical draining and containing, equipment disassembly, and crating and packaging, rigging, as well as costs for packaging materials, crates, carts, chemical shipping containers, and so on.

On completion of the maintenance budget worksheet for each type of equipment, the results of each sheet can be rolled up to the department main budget summary sheet (Figure 5.3). The summary sheet also includes department-level cost categories such as headcount payroll costs, supplemental resource costs, and general spending to keep the department operating. Some categories shown in the individual equipment worksheets, such as training, tools, software, and CI projects, also require data entry on the summary sheet as some of these expenses are not equipment specific.

Payroll costs typically include base pay, overtime pay, benefits, and bonuses. Costs for supplemental resources are related to costs spent on consulting, hiring temporary workers, and hiring contractors. Temporary workers are generally for work augmentation, which means hiring additional help to perform job functions that exist internally. Contractors are normally hired to perform jobs that require skills lacking in internal personnel. The general spending category includes all the regular operating expenses. The examples are (1) safety supplies such as safety glasses, shoes, and labels; (2) communication spending such as landline phone bills, cell phone bills, and Internet provider charges; (3) team events such as departmental team-building events as well as functions with customers and suppliers; and (4) miscellaneous supplies such as cabinets, chairs, workbenches, office supplies, and computing supplies. Depending on the nature of the business and the financial rules of the company, some of the items in this category may be in the company general spending budget instead of the departmental budget.

The budgeting approach described in this section is an example of bottom-up budgeting, which serves as a general guidance in maintenance budgeting. Many companies have standard templates and practices for generating departmental budgets, so modification from the example worksheets given is required to meet the specific needs of a particular company. Even if the company does not require such a detailed budget plan, it is a good practice to have the details planned so the budget is justified and tracked easily. In addition, many managers also keep track of budget changes from year to year and generate run rate indicators to demonstrate productivity and efficiency improvements. These indicators are discussed in the next chapter. Submitting the budget plan along with these indicators significantly increases the probability of budget approval from upper management.

Tactical Planning in Maintenance

While strategic planning is high-level planning to ensure that the department's general direction and capabilities meet the business requirements, tactical planning involves low-level management activities to ensure that all the daily activities are carried out efficiently. It consists of three major areas of activities: task assignments and scheduling, resource scheduling, and process planning. The tactical planning activities are traditionally done by maintenance managers, shift supervisors, and planners. With the development of advanced maintenance management software, many of the activities can be effectively done by computers.

Maintenance tasks need to be assigned to the right technicians with the right skills. Traditionally, maintenance shift supervisors review incoming work orders and scheduled maintenance work, then assign work to individual technicians. Currently, many companies have maintenance management systems set up to assign work orders and schedule work automatically based on the technician databases set up in the systems.

A key task assignment activity is setting the right priority for the tasks. Since a large portion of maintenance work is unscheduled repairs and assists, and the maintenance headcount is planned based on average downtime, there will be situations for which resources are not enough to cover the work. In these cases, the right priorities need to be set to guide the technicians on what equipment is to be worked on first and what equipment will need to wait. The priorities should be planned ahead and typically negotiated with the equipment users. Figure 5.4 shows a priority sheet example that is communicated to the entire department so management does not have to make the decision on the spot when resources

Priority	Equipment Type	Equipment ID	Location	Resources	Activity
1	Equipment1	System#A11	Building#1 RM101 A2	EEW Certified, Tech 4+	Repair
2	Equipment1	System#A12	Building#2 RM102 A4	EEW Certified, Tech 4+	Repair
3	Equipment2	All	Building#3 B2-E8	Tech 2+	Setup
4	Equipment3	System#C31	Building#3 E9	Tech 2+	PM
5	Equipment3	System#C32	Building#3 E10	Tech 3+	Repair
6	Equipment4	System#D41	Building#4 RM202	Laser Certified, Tech 3+	Repair
7	Equipment5	System#E51	Building#3 F2	Tech 3+	Repair
8	Equipment3	System#C33	Building#3 F10	Tech 5	Upgrade
9	Equipment5	System#E52	Building#3 F4	Tech 3+	Repair
10	Equipment5	System#E53	Building#3 F6	Tech 2+	PM

FIGURE 5.4
An equipment priority worksheet example template.

are constrained. It details the equipment priority, locations, activities, and resources required to carry out the activities. The priority sheet also needs to be updated regularly—sometimes monthly, weekly, daily, or even every shift—based on the nature of the operations. Color codes may also be used to highlight the priorities and gain more user attention on key items.

Finally, planned maintenance tasks such as PM and upgrades need to be scheduled properly. They are typically planned around production schedules and often are scheduled in off-peak operational hours or on the weekends. Other maintenance activities such as upgrades, reconfigurations, and CI projects should be planned around a production schedule to minimize the impact on the users. If possible, these activities should also be performed together with PM to minimize downtime. Major equipment upgrades and PM that require vendors to perform should be scheduled during periods of low equipment utilization but avoid doing them during off-hours, which typically means additional charges from the vendor.

Resource scheduling means having the right resources at the right time to support the maintenance activities. Main maintenance resources are the workers, tools, and materials. Maintenance personnel scheduling must be done based on production schedules and in conjunction with task scheduling. Assigning maintenance techs with the right skill sets to perform the corresponding tasks is also important. Mismatching skills and tasks may lead to safety incidents, equipment damage, or prolonged troubleshooting time. The maintenance department manager and supervisors should maintain a file on all maintenance technicians with indications of their equipment certifications. More details are discussed in the training and people development section.

Tools and materials planning is another major area in tactical maintenance resources planning. Using the right tools to perform the right tasks not only reduces safety and ergonomic issues but also speeds up the repairs and PM. Making all the necessary materials available for the maintenance tasks is also critical in reducing equipment downtime. Managing maintenance tools and materials is elucidated in greater detail in the inventory management section.

The last major tactical planning area is process planning. Maintenance is the business that deals with excursions. To reduce firefighting and stress, proper procedures are developed to anticipate issues and to define a process to solve them. Specific actions may not be clear in solving difficult equipment hard-down issues, but specific processes can be planned to guide the maintenance workers to identify the root cause of the issues

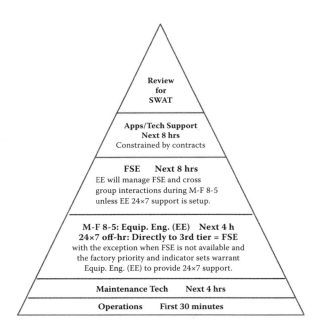

FIGURE 5.5
An example of escalation structure.

and to solve them. This process-planning tactic is sometimes called excursion management.

The procedures generated in the process are standard operating procedures (SOPs), response flow charts (RFCs), and excursion escalation procedures (EEPs). SOPs and RFCs are typically equipment specific and developed by the internal equipment experts by consulting the equipment manufacturers. SOPs generally have more text and read like equipment menus. RFCs are typically in graphical form with standard shades and arrows that demonstrate the series of actions and decisions steps. EEPs provide guidance on when to take the next level of actions.

Figure 5.5 demonstrates an example escalation structure that starts from the first-level operations response to maintenance techs, to equipment engineering, to vendor field service engineering, to vendor application engineering and technical support, and to review for possible SWAT. The term *SWAT* is borrowed from the abbreviated term for a military term, special weapons and tactics, as the ultimate problem resolution task force that typically works around the clock until the problem is solved. It includes everyone in the escalation path as well as senior management from both the factory and the vendor. The graphical display is a part of the escalation procedure, which includes a detailed description of the conditions as well as contact information for each level of support.

While the maintenance department strategic plan may not need to be communicated to all employees, depending on the company's corporate culture, of course, the maintenance tactical plans must be communicated to all maintenance personnel. In fact, a system needs to be developed to ensure that the tactical plan items are available, updated to reflect the latest equipment and personnel changes, and are followed with discipline.

TRAINING AND PEOPLE DEVELOPMENT

Maintenance personnel need the right skill sets to repair and keep the equipment in proper condition. They are like the doctors and nurses of equipment. Training generally refers to obtaining the needed skills for the current job functions, and development is aimed at the future growth of the employees. Maintenance training and development generally can be grouped into three main areas: (1) equipment-specific training on the operation and maintenance of an individual type of equipment; (2) field-specific training and development on the fundamental knowledge of a profession or trade, such as for mechanics and electricians; and (3) general training and development on safety, problem solving, business processes, maintenance management, time management, inventory management, project management, and so on. Before jumping right into setting up training and development programs for each individual in the maintenance department, the manager must formulate a training and development plan based on the needs of the department and employees. Figure 5.6 illustrates a general process used to develop and implement the training and development plan for the maintenance personnel.

Determining the development direction is the first step in formulating the training and development plan. The maintenance department is facing two distinctive directions in developing its personnel: universal tech or specialist. Universal tech means to train and develop everyone in the department so that they have the ability to respond to any equipment issues in the factory. Under this approach, the maintenance techs are trained in a wide range of skills. The advantages of the universal tech are faster customer responses, more resource flexibility, and less waiting for tech time. Equipment users can grab any maintenance tech to help

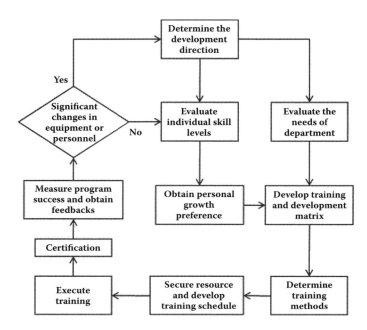

FIGURE 5.6
A general process for maintenance training and development.

them when they have equipment problems, whether the problems are mechanical, electrical, electronic, or with software. Task assignment and scheduling are made easier as a bigger pool of resources is available to various maintenance works. Maintenance technicians are not separated into technical areas, so there is no need to wait for mechanical techs to repair mechanical failures, electronic techs to solve electronic issues, and so on. However, the disadvantage of this approach is that learning a wide range of skills would generally have trade-offs in the in-depth knowledge to solve hard-down difficult equipment issues. Without the in-depth skills in the department, vendor dependence is high, which may lead to expensive contracts and longer downtime on more difficult repair incidents with added logistics to get the vendor FSE on site.

The specialist approach aims to create content experts in each area by training and developing individual maintenance technicians with skills that are close or even equivalent to the proficiency of vendor FSEs. The advantages of the specialist approach are less vendor dependency, more professional department image, as well as greater opportunities for equipment efficiency improvements. With in-house content experts, the factory can reduce contract and extended warranty requirements to save costs.

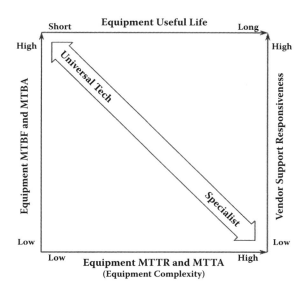

FIGURE 5.7
A training and development direction decision model.

The internal experts also allow the setup of an around-the-clock support structure without expensive costs. With many highly capable technical experts, the department will certainly exhibit a professional image, garnering respect from the user community. The specialists also enable the factory to tailor the equipment to meet its specific operational requirements for efficiency gains without worry about intellectual property (IP) leaks and expensive non-recurring engineering (NRE) costs.

The decision of the training and development direction should be situational and determined based on the equipment characteristics and business environment. Figure 5.7 is a decision model for determining which direction is a better fit. When equipment MTTR and MTTA are high, it generally means equipment complexity is high, and in-depth skills are lacking to solve equipment issues in a timely manner. Therefore, having specialists on site is desirable. When equipment MTBF and MTBA are low, equipment downtime frequency is high; having specialists is also preferred so equipment reliability can be improved. Examining this in the opposite direction, low frequency of equipment downtime cannot justify the effort in training specialists. If the specialists do not have the opportunities to use and practice their skills, they will lose proficiency over time. When equipment useful life is short, the introduction rate for new equipment is high. The equipment does not stay long enough for the maintenance personnel to develop expertise to become specialists. When the vendor support responsiveness is low, that means the vendors do not have infrastructures in place to respond to equipment support requests in a timely manner. Since vendor services are not dependable, internal specialists are needed for self-sufficiency in resolving equipment issues.

Other than the four dimensions shown in the model, aspects such as requirements from customers and management as well as the availability of skillful personnel could also be constrained factors in pursuing either the universal tech or specialist approach. In some cases, hybrid approaches of universal tech and specialist can be implemented by (1) selecting a few highly skilled workers to form a specialist team for difficult issues while most of the maintenance personnel remain as universal techs or (2) requiring each maintenance technician to become a specialist on one machine while being a universal tech on the rest of the equipment. The training and development direction choice warrants careful consideration as it determines the maintenance department's support model.

Once the direction is clear, the next two steps are evaluating the needs of the department and the skill level of each individual. These two steps can be done in parallel. The purpose of evaluating the training needs of the department is to generate a list of areas and skills required to maintain the equipment effectively as well as determine the number of technicians trained to cover each type of equipment. Evaluating the current skill level of each individual in the department can be done by self-assessment, then management assessment. An example of an assessment form is illustrated in Figure 5.8. After completing the skill assessment, the next step is to obtain a personal growth preference from individuals. Asking the individuals what training and development they would like to pursue empowers them to participate in their own career development so they are more likely to commit and follow through with their own decisions. Empowered employees will have better morale and lower absenteeism.

By consolidating the department training needs, the skill levels of individuals, as well as the individual's preference in development, a training matrix can be generated, as demonstrated in Figure 5.9. The top row of the matrix lists the training classes. This list should include classes from all three areas mentioned: equipment-specific training, profession field-specific training, and general training. The first left column lists all the maintenance technicians. The second column lists their working shifts, which highlights any imbalance in the number of trained workers in a particular

TECHNICIAN _____   DATE _____
SHIFT # _____

INSTRUCTIONS: Enter a number from 0 to 4 along the right side of the equipment list below. This number should reflect your current repair ability level to the corresponding equipment. Use the below definition criteria to properly identify your skill level.

NOTE: The requested data will be summarized to evaluate the training skill set of the department as a whole. This information will be used to help determine out training requirements. It is not a reflection of your contribution level or performance. Please be as accurate as possible when assigning skill levels to the equipment types.

Equipment Type	0-4
Equipment1	
Equipment2	
Equipment3	
Equipment4	
Equipment5	
Equipment6	
Equipment7	
Equipment8	
Equipment9	
Equipment10	
Equipment11	
Equipment12	
Equipment13	
Equipment14	
Equipment15	

DEFINITION CRITERIA

0 = No equipment exposure/support

1 = Minor debug/repair. You can perform setups and assist other techs.

2 = First level repair: You can run system diagnostics and replace parts that it flags. Typically you will have completed vendor training or equivalent. For mechanical equipment you can perform basic adjustments.

3 = Very proficient at equipment repair. You are able to go beyond diagnostic evaluation and repair the equipment based on your past experiences and thorough knowledge of its operation.

4 = Equipment expert. You are able to fully comprehend all aspects of the equipment. You are typically sought after by users to resolve major problems or are able to modify or redesign the equipment for reliability and productivity improvements.

FIGURE 5.8
A training assessment example form.

area between shifts. In the example training matrix, an *x* means the completion of a particular training by a technician. It can be replaced by the training completion date if that level of detail is demanded, such as when annual refresher training is required. Therefore, this training matrix is not only a planning tool but also a tracking tool. On a side note, the training matrix can be used to increase the employees' initiative to improve their skills. By making this matrix available to everyone in the department, technicians can see what skills their peers possess and may spark friendly competitions to earn more check marks. Also, when an employee sees that his or her skill sets are far behind his or her peers, the employee should be alerted that employability in the department may be in jeopardy if a layoff is coming. Hence, the employee would be more motivated to improve the skills to remain employed.

After the training matrix is in place, the next step is determining training methods. For equipment-specific training, the choices are off-site vendor training, in-house instructor-led training, and peer-to-peer cross-training. Off-site vendor training can be expensive, but the students get the most knowledge without the interference of regular job duties. Another option is to pay for the vendor instructor to conduct

	Equipment1	Equipment2	Equipment3	Equipment4	Equipment5	Equipment6	Equipment7	Equipment8	Equipment9	Equipment10	Equipment11	Equipment12	Equipment13	Equipment14	Equipment15	
TIME MANAGEMENT	x	x	x	x	x	x	x	x	x	x	x	x	x	x	x	
PROBLEM-SOLVING TECHNIQUES	x	x	x	x	x	x	x	x	x	x	x	x	x	x	x	
HYDRAULIC SYSTEMS		x	x					x								
MECHATRONICS		x	x					x								
ELECTRICAL MEASUREMENTS			·	x		x	TBD		x			TBD	x			
DC CIRCUITS	x			x		x	TBD		x		x	x	TBD			
CAD DRAWING	x	x		x		x	x	x	x		x	x			x	
HAZARDOUS MATERIALS	x	x	x	x	x	x	x	x	x	x	x	x	x	x	x	
ELECTRICAL SAFETY	x	x	x	x	x	x	x	x	x	x	x	x	x	x	x	
EQUIPMENT3 PROGRAMMING	x	x	x				x				x					
EQUIPMENT3 MAINTENANCE LEVEL3																
EQUIPMENT3 MAINTENANCE LEVEL2	x	x	10/10	x							10/10	x	x			MMYY = DATE SCHEDULED
EQUIPMENT3 MAINTENANCE LEVEL1	x	x	x	x				x			x	x	x			
EQUIPMENT3 PM	x	x	x	x			x				x	x	x			
EQUIPMENT3 SETUP	x	x	x	x	x	x										
EQUIPMENT3 OPERATIONS	x	x	x	x	x	x	x	x	x							
EQUIPMENT2 APPLICATIONS							x	x	x					x		
EQUIPMENT2 MAINTENANCE	x	x	x	x			x	x	x	x				x		
EQUIPMENT2 PM	x	x	x	x			x	x	x	x	11/10			x		TBD = TO BE DETERMINED
EQUIPMENT2 SETUP	x	x	x	x	11/10		x	x	x	x		x		x		
EQUIPMENT2 OPERATIONS		x		x	x	x		x	x	x		x	x		x	
EQUIPMENT1 PROGRAMMING				x	x	x										
EQUIPMENT1 MAINTENANCE LEVEL2		x	x	x	x	x	TBD	x	x			x			x	
EQUIPMENT1 MAINTENANCE LEVEL1	x	x	x	x	x	x	x	x	x	TBD	x				x	
EQUIPMENT1 PM	x	x	x	x	x	x	x	x	x	x	x	x	x			LEGEND X = COMPLETED
EQUIPMENT1 SETUP	x	x	x	x	x	x	x	x	x	x	x	x	x	x	x	
EQUIPMENT1 OPERATIONS			x	x	x	x	x			x	x				x	
SHIFT	#	#	#	#	#	#	#	#	#	#	#	#	#	#	#	
TECHNICIAN	Equipment1	Equipment2	Equipment3	Equipment4	Equipment5	Equipment6	Equipment7	Equipment8	Equipment9	Equipment10	Equipment11	Equipment12	Equipment13	Equipment14	Equipment15	

FIGURE 5.9

A training matrix example template.

the class in-house. It costs less but typically requires scheduling the equipment to be off line for training purposes, which interrupts operations. Also, maintenance work situations, such as an outburst of down incidents, may interfere with the class. The most widely used training method is on-the-job peer-to-peer cross-training. It is the least expensive but requires internal experts with trainer skills. Not all equipment specialists are good trainers. For the companies that use peer-to-peer cross-training as their main training method, training and coaching skills should be one of the training areas in the training matrix for the senior technicians. The advancement of computer and Internet technologies makes computer-based training (CBT) and Web-based training (WBT) popular. However, there is minimal application of CBT and WBT in equipment training since hands-on training and physical presence around the equipment are essential.

For the professional field-specific training and general training, companies have more choices. If there are a large number of employees who need to be trained, companies can set up partnership programs with local universities and colleges to have special classes. Otherwise, tuition reimbursement programs can be set up to allow individuals to take classes as open-university students. Field-specific training consultants can be brought on site to conduct the classes as well. Internal experts can also be the instructors of these classes. Many of these classes can now be conducted using long-distance learning tools such as CBT, WBT, teleconferencing, and videoconferencing.

Some companies emphasize self-learning. Computerized expert systems (also called knowledge-based systems) are developed and implemented to capture such knowledge as general engineering methods (GEMs) and best-known methods (BKMs). Employees are expected to use the systems to solve work problems. If the knowledge for a particular problem does not exist in the system, the employee is expected to contribute his or her own finding in solving the problem to the system so others can learn from it in the future. It becomes a living system that evolves over time and updates recurrently.

The next step in the training and development process is training arrangement and logistics, such as securing resources and developing training schedules. Large companies have training departments or dedicated training coordinators that provide such services. If such a function does not exist, the maintenance manager needs to manage it or delegate it to a person in the department with good organizational and communication skills. Hiring training consulting firms to manage and provide the training is another option, but the proper cost needs to be included in the budget plan.

With the training arrangements completed, the next step is conducting the training and executing the development programs. To ensure the training is conducted with optimal effectiveness, managers should schedule resources to cover the operations for those who are in training and avoid interruptions caused by the trainees' normal job duties. After the training is completed, the certification step must be followed to ensure that the trainees master the skills learned. Many classes include exit exams that can be considered as formal certification, but for those classes that do not, a process should be set up to have the trainees demonstrate the skills to management or senior technicians. Certifications must also be documented and dated. A certification matrix, similar to the training matrix shown in Figure 5.9, can be used to track certifications of all the maintenance personnel. The certification matrix typically has fewer columns than the training matrix as a certification on an area or equipment may include the completion of several classes.

The current business environment demands that an organization must be continuously learning and developing. There is no exception for the maintenance organization. Training and development is a revolving process. To set up the next training and development program, management must measure the success of each program after its completion and request feedback from the participants. Based on the program evaluations, as well as any recent business environment shifts, such as equipment and personnel changes, new training and development plans can be formulated again to advance the organization's capability to resolve upcoming equipment issues.

CUSTOMER SERVICES AND MANAGEMENT

Maintenance is a service organization with internal customers, the equipment users. However, many maintenance technicians in traditional organizations have an ego like a typical auto mechanic. They feel that equipment users are coming to them with problems and are at their mercy. Often, they feel that the equipment issues are caused by the users, especially if the users are operators who have lower skills and job grade levels than they do. Thus, they either begin to accuse the users or unintentionally show impatient attitudes that make the users feel uncomfortable and stupid. It is human nature to feel good when being valued and seen as imperative, so many people like the fact that others are begging for their help, which makes them feel important. If management does nothing, over time these behaviors would develop and hurt the organization by creating barriers between the maintenance department and the user community. The equipment users would avoid contacting the maintenance group for help unless absolutely necessary, therefore leading to additional downtime and inefficiency in equipment utilization.

Customer orientation should be a constant focus for maintenance management. It is recommended to add a customer service course in the training and developmental matrix for all technicians. Many consulting firms offer such classes, but most are oriented toward the retail and sales industries. Managers should pursue the possibility of tailoring the classes for the maintenance operation. In fact, maintenance organizations should practice customer management rather than just pure customer service.

Customer service is based on the philosophy that "the customer is always right." Associates are told never to challenge the customers. The key ingredients in customer service are (1) responding to customers' requests in a timely manner; (2) listening to customer complaints attentively with a positive attitude (smiling); (3) acknowledging and agreeing with customers' points of view; (4) giving sincere and blameless apologies; and (5) resolving customers' issues quickly and in an error-free manner. In practical training, the students are trained never to say "no" to the customers and instead use such recommended sentences as "I understand your frustration"; "That is true"; "I regret that this has happened"; or "I am sorry you are having this trouble." Implementing a customer service approach in maintenance organizations can certainly change the arrogant attitude of the technicians and promote a customer-oriented image of the department. However, due to the dynamics in equipment management, totally turning the table to the other side is not necessarily good for the organization. Operations are under great pressure to produce outputs, and equipment issues can be the best excuses for not meeting the required output. As customer service is rather reactive and emphasizes solving the problem from the receiving end of the issues, taking this approach too far would make the maintenance department a dumping ground for issues.

Customer management is a more appropriate approach for maintenance organizations as it creates a two-way cooperative environment for both parties to solve problems with shared professionalism and responsibilities. Customer management entails (1) setting mutual expectations with customers, (2) monitoring the user's operating environment and behaviors for optimal situational control, (3) educating customers to avoid unreasonable requests, and (4) proactively communicating to customers on equipment conditions that may lead to issues.

First, maintenance managers should obtain equipment users' expectations and have a full understanding of the achievability of such expectations. Based on careful study of user expectations, managers then negotiate with the equipment users and build a set of expectations for the customer in return. An example of such expectation is to require the users to flag the equipment failure in real time using a CMMS and provide a detailed description of the issue. Such expectations typically allow cooperation between the parties to meet the mutual objectives.

Second, maintenance managers need to keep track of the production schedule and the usage model for the equipment. It is like a weather forecast that provides warning when a storm is coming. If possible, implementing a system to monitor how users utilize the equipment can provide vital data for problem solving and for customer education, which is discussed next. As mentioned many times, maintenance is the business of dealing with unpredictable events. People in general do not react well to unfavorable surprises. Therefore, maintenance managers should also monitor customers' behaviors and sensitive areas (hot buttons) to anticipate customers' reactions on equipment excursions that will have an impact on their planned activities. Such information will allow better control over these unforeseen situations and reduce frustrations for both sides.

Third, customer education is an important part of customer management. Since the customers are not experts in the equipment (and that is why they need help), they often do not understand the difficult factors in resolving an equipment problem or performing a conversion. For example, they may miss important information that can significantly speed up the repairs or they may make setup requests that require enormous efforts but can be easily avoided by small changes in operations processes. The equipment users have no intention of making the maintenance jobs more difficult, but if they are not educated on how the equipment should be used, what data are important for troubleshooting, or how long it takes for a certain activity, the maintenance personnel will certainly have a hard time completing the tasks.

Fourth, rather than reactively waiting for the equipment users to come with issues, the maintenance manager and technicians should proactively provide equipment status and condition reports to the user community regularly. The reports should include performance indicators for each individual machine to advise the user what the top failures associated with the machines are and which machine has more issues. Any condition that may lead to equipment usage interruption should be communicated to the users as well. The information allows the equipment users to plan their activities better and adjust their usage model to best utilize each machine.

VENDOR, SUPPLIER, AND CONTRACT MANAGEMENT

Vendor relations and contract management are essential to the success of the maintenance operations as maintenance personnel regularly rely on the support of equipment vendors and suppliers for spare parts and tools. The formal relationships with the vendors and suppliers are defined by the contracts between the companies. However, establishing a strong relationship with the vendors and suppliers can reduce the wait time for support requests and speed up part deliveries, as well as facilitate additional services beyond contractual obligations.

To manage the vendor and supplier relations effectively, maintenance managers must have a full understanding of the contracts. Generally, the first contract between the parties is the equipment purchase agreement. In most companies, maintenance managers are typically not involved with the negotiation of the purchase agreement. Therefore, maintenance managers must make the effort to seek out the right people to obtain the agreement and then study it carefully. Table 5.1 shows the table of contents of a purchase agreement example. It is filled with legal language, often quite lengthy. Of all the sections, maintenance managers should pay most attention to information on warranty, equipment reliability, spares, and services.

New equipment generally comes with a warranty; however, maintenance managers should not assume that it will cover everything. Typical topics found in the warranty section are warranty conditions, warranty period, parts coverage, labor coverage, support hours, response time, and responsibilities of both parties. First, managers must pay special attention to the warranty conditions since these are the requirements that keep the warranty valid. If the equipment is misused or the owner does not take proper care as outlined in the purchase agreement, the warranty is voided, and the repair will be done at the buyer's expense. Next, all the coverage

TABLE 5.1

Example Purchase Agreement

Contents

Part I: General Terms and Conditions

1.1 Definitions
1.2 Terms of Agreement
1.3 Termination for Convenience
1.4 Contingencies
1.5 Packing and Shipment
1.6 Ownership and Responsibilities
1.7 Confidentiality and Publicity
1.8 Intellectual Property Indemnity
1.9 Customs Clearance
1.10 Compliance with Law and Rules
1.11 Insurance
1.12 Indemnification
1.13 Retention and Audits
1.14 Independent Contractor
1.15 Security
1.16 Merger, Modification, and Waiver
1.17 Assignment

Part II: Equipment Terms and Conditions

2.1 Product Configurations and Specifications
2.2 Pricing
2.3 Invoicing and Payment
2.4 Delivery, Releases, and Scheduling
2.5 Acceptance
2.6 Warranty
2.7 Modifications and Upgrades
2.8 Change Control
2.9 Training
2.10 Safety Review and Notification
2.11 Drawings and Documentation
2.12 Equipment Reliability

Part III: Spare Terms and Conditions

3.1 Spare Definitions
3.2 Spare Terms
3.3 Spare Pricing
3.4 Spare Consumption
3.5 Spare Lead Times

TABLE 5.1 *(Continued)*

Example Purchase Agreement

Contents
Part IV: Service Terms and Conditions
4.1 Applicability
4.2 Service Pricing
4.3 Equipment Predelivery and Install
4.4 Field Service Support
4.5 Continuous Improvements
4.6 Escalation
4.7 Other Seller Responsibilities
4.8 Buyer Responsibilities

details must be fully understood. The parts coverage period may not be the same as the labor coverage period. Some parts may not be covered at all. In addition, a warranty typically covers repairs made only Monday to Friday, 8 a.m. to 5 p.m., which is not sufficient to cover operations that run 24 hours a day and 7 days a week. Also, most warranties have a response time of next business day, which may not meet the needs of a critical operation. Therefore, maintenance managers plan, budget, and negotiate for extra coverage beyond the warranty to meet the operational needs. The warranty section also outlines the responsibilities for both parties during regular equipment use, PM, and in the case of equipment failures. Maintenance managers should fully understand the expectations from the vendor in terms of keeping the warranty valid and the logistics in exercising the warranty options.

The equipment reliability section in the purchase agreement states the equipment performance goals, which allows the maintenance managers to determine if the equipment is performing within an acceptable reliability range. Benchmarking the actual equipment performance against the goals stated in the purchase agreement is one of the key bases for formulating vendor management tactics as it enables maintenance managers to determine when to involve the vendors, how much negotiation power they have, and how much pressure should be applied to hold the vendors accountable for the issues. Not all the purchase agreements have penalty clauses associated with not meeting performance goals. If possible, maintenance managers should participate in the negotiation of the purchase agreement, and one of the objectives is to push the penalty causes when equipment is performing under the goals. The penalty to the vendors could be monetary or simply extending the warranty, without charge, for the period that performance goals are not met.

The spares section lists the spare items that are recommended to stock for quicker equipment repairs. It also defines the options for obtaining the spares as well as the pricing for the spares. In addition, it describes how to replenish spares when they are consumed as well as the lead time to get them. For repairable items, instructions are given regarding how to handle the exchanged parts as well as getting them back into the spare inventory. Spare management is discussed in detail in the next section, but the negotiation of spare terms and conditions is an important part of the vendor interface, so it is worth mentioning here. Maintenance managers should push for documenting the following vendor responsibilities in this section: (1) provide regular spare consumption report, (2) agree to take actions to reduce spare consumption by a certain percentage each year, (3) credit or refund the cost of spares that are not used for a prolonged period of time, (4) offer a discount on parts that fail frequently, and (5) offer emergency out-of-stock services for quick delivery.

Finally, maintenance managers must read word by word in the service terms and conditions section, which provides the details of which services are offered by the vendor at what price. These services typically include equipment installs, upgrades, factory change orders (FCOs), engineering change orders (ECOs), as well as field engineering on-call and on-site supports. The escalation path is also defined for issues that need additional support beyond FSEs. In addition, this section clarifies the roles and responsibilities for both parties to ensure the services are initiated and carried out properly. Examples of vendor responsibilities are (1) providing qualified FSEs with minimal personnel assignment changes during the contract period; (2) complying with the buyer's company policies, safety and environmental procedures, and IP guidelines; (3) offering work schedules that correspond with a buyer's site-specific holidays and factory schedule, including support of scheduled facility maintenance shutdowns; (4) providing the buyer with equipment-specific tools and standard tool kits; and (5) providing equipment failure data and analysis associated with the purchased equipment model in the field, not just in the buyer's company if requested. Since it is a mutual agreement, the equipment buyer's responsibilities are also documented in detail. Typical buyer responsibilities are providing (1) access to equipment for PM and repairs;

(2) factory contacts to define priorities and assist in resolving personnel issues; (3) access to required facility, product, and equipment performance documents for equipment repair purposes; (4) a schedule of holidays and planned shutdowns; (5) Internet and network access for communication; and (6) a work area or office for field service personnel.

Other purchase agreement sections, such as safety and training, also need to be comprehended. Understanding the safety precautions for the equipment enables the development of safety measures to reduce accidents. Checking the training section can provide information on what training classes are needed and offered for maintenance personnel and if training credits are available to reduce the maintenance training budget.

In most companies, equipment purchase decisions are not in the hands of maintenance managers, so purchase agreements are most likely done without maintenance managers' involvement, but they need to work with the vendors under the terms defined by the purchase agreement. This may lead to situations in which the maintenance managers are powerless when requesting services from the vendor. At the corporate level, it is recommended that maintenance managers are included as part of the equipment purchase decision team. If this top-down approach is not typical or realistic at the company, the maintenance manager should take the initiative to interface proactively with the vendor as early as possible to establish a stronger relation.

Once the warranty clause in the purchase agreement expires, maintenance managers most likely own the decisions on the extended warranty contracts or time and materials contracts, so they start to have more negotiating power in working with the vendor. Any difficult factors encountered during the warranty period should be taken into consideration for negotiating the terms of the extended warranty contracts and services. If possible, managers should seek third-party support on the equipment to maximize the negotiating power and minimize costs. It is similar to taking a car to a general auto shop rather than going to the dealer for services. The more options are available, the more power the maintenance managers have. Developing internal specialists is another option that can reduce vendor dependence and balance the power to establish a shared partnership with the vendor.

Maintenance managers also need to establish and manage the relationships with the suppliers for spares, materials, and tools. Unlike the relationship with the equipment manufacturers, the relationships with these suppliers are largely within the maintenance managers' control since they are typically initiated by the maintenance department. A formal supplier certification process is highly recommended so all the suppliers are selected under the same criteria. Also, a second source backup supplier should be identified for each supplier selected. In addition, the selection should not be done just once and maintained for years even though it saves administrative effort. A regular review or audit should be in place to ensure the services and prices provided by the selected suppliers are at the optimal level.

INVENTORY MANAGEMENT

Maintenance inventory includes spares, tools, and materials. There is an increasing trend for equipment purchasing cost, and the same is true for the maintenance inventory cost. Managing inventory is an important logistic task for maintenance operations, and it deserves the same scrutiny as managing the factory's raw materials, work in process, and finished goods inventory. Many companies apply the same inventory management approaches to manage both production and maintenance inventories. While some inventory-managing methods can be shared, maintenance inventory, especially for spare parts, has some unique characteristics that require special attention and consideration.

Spare parts typically account for a significantly large portion of the maintenance inventory. Obtaining spares on site can greatly reduce equipment downtime. The initial task in setting up an on-site spare program is determining the spare part list with the minimum and maximum level for each item on the list. The minimum level is the safety stocking level and the reordering point for the items. The maximum level is in place to limit the costs for obtaining and maintaining the spares. Spares that exceed the maximum count should be transferred to other factories or sold back to the vendors. Ideally, the on-site spare list and its minimum/maximum counts should be determined based on the usage, which can be derived from the equipment failure data. However, the spare list is needed when a new machine is introduced to the factory, and the usage data may not be available at the time. In this case, the spare list is based on an equipment manufacturer's recommendation, but maintenance managers must understand that the equipment manufacturer is generating the list from its perspective and for its benefit. Assumptions and failure data that are used to generate the recommended spare list should be obtained from the equipment manufacturers. A different equipment user model may alter the assumptions and drive for different spare requirements. After the equipment is installed and operational, spare usage and failure data should be tracked and reviewed regularly. The spare lists and the minimum/maximum levels should be modified if needed.

A spare management decision that maintenance managers may face is purchase versus consignment; essentially, it is a choice between buying and renting. Many equipment manufacturers offer a consignment program on their spares, which puts spares at the equipment buyers' location for a fee. The benefit of purchased spares is that the spares are totally owned by the buyer and can be used in any way desired by the company. The disadvantages are high cost and inflexibility in modifying the inventory. The money is wasted for purchasing spares that end up not being used. Negotiation with the manufacturers to buy back the unused spares is an option, but it often means taking a loss. If equipment life and spare shelf life are short, which is a trend for many modern machines, the cost factor becomes even more significant. Therefore, more companies negotiate and choose the consignment program rather than purchase spares, especially for high-value parts. When choosing the consignment spare option, maintenance managers must ensure that the company is carrying additional insurance to cover the full replacement costs of the spares in case of any disaster. Also, the consignment spares are not owned by the company and may not be used for equipment development, upgrading, and characterization purposes. Part damage due to such activities may not be covered, and a full purchase price of the spare must be paid to the vendor in these cases.

Maintenance inventory stockroom management requires careful planning, detailed tracking, and frequent auditing. A typical maintenance inventory has thousands of parts, so systematic stocking and tracking methods are absolutely essential. First, an internal part naming and numbering system must be developed. Part manufacturers and vendors have their own numbering conventions with different letter-number combinations and lengths. Hence, just using the vendors' part numbers would be confusing and might even present duplicate or similar part numbers that represent totally different categories of parts. The internal naming and numbering convention must be consistent and intuitive in representing the maintenance inventory so that the maintenance workers can indentify, just by looking at the part number, the type of parts for which type of machine and even stocking location.

Second, the physical layout of the stockroom must be designed systematically for easy and quick access to stock, audit, and find parts. Similar to the library system, the stocking system should be based on the internal part numbers, which can reflect the locations and categories of the parts. Maintenance managers and planners must pay attention to not only how to organize the parts but also how to store the parts for effective inventory control. For instance, a two-bin system can be implemented to prompt for reordering and avoid an out-of-stock situation [12]. Instead of putting all the identical parts in one big bin, parts are separated into two bins, with the bottom bin containing the minimum safety stocking level while the top bin holds the rest. When the parts are used in the top bin, it visually alerts the maintenance workers to reorder the parts while the equipment can still be sustained with the parts in the bottom bin as a safety buffer for the ordering lead time.

Third, parts should be classified to ensure the right inventory control focus and efforts on the right parts for cost saving and efficiency. Not all items in the maintenance inventory require the same attention. A popular method in inventory management called the ABC analysis is highly recommended. It is a process of classifying the entire inventory into three classes based on the part values [12]. Class A has the high-value parts, and it typically correlates with the 20–80 rule: 20% of the total items accounts for 80% of the value and usage. Items in this class are also known as the critical parts. Class B has the midvalue parts, which typically account for 30% of the total inventory. The remaining 50% of the inventory is in class C. The objective of the ABC analysis is to identify the inventory classes so specific tactics can be developed to manage each class effectively. For class A parts, a system is needed to monitor usage closely. It requires more attention and vendor management efforts to reduce the stocking level to save costs while ensuring these parts are handled with quick delivery and repair turnaround times. For class C parts, continuous monitoring is not needed, and a periodic check, such as monthly or quarterly, would be sufficient. Ordering these parts in bulk should be considered as it can provide discounts for cost saving. Class B parts fall in the intermediate level between class A and class C for frequency of control.

The application of computing technology to inventory management is common in most industries and companies. It is easier to manage the maintenance inventory with a computerized inventory control system. When choosing a system, maintenance managers often face several choices: (1) the company's production inventory management system, (2) a maintenance management system that has a spare management module, and (3) a stand-alone maintenance-specific inventory management system.

Using the company's production inventory system for managing maintenance parts can save administrative costs and overhead for the maintenance department. The major drawback is that maintenance inventory has different characteristics compared to production materials. For example, a repairable spare needs to be logged in and out of the systems with the need for traceability of the same part. Nevertheless, maintenance managers should review the company's production inventory-tracking system with an open mind to determine if the system can be used to accommodate spare management needs. Some production inventory systems actually have separate modules tailored differently for managing raw production materials, work-in-progress materials, and finished goods. Adding a spare management module may not be a difficult task.

The next option is through selecting or developing a maintenance management system that has a maintenance inventory management function built in. Since a CMMS is almost indispensable for effectively managing the maintenance function, adding on the inventory management module would not require too many additional resources. The primary function of most maintenance management systems is to keep track of equipment failures, downtime, and work orders. Integrating the equipment failure tracking with spare management certainly provides additional benefits as the failed parts history can be tied to each machine failure. The functions of CMMSs are discussed further in Chapter 7.

The last alternative is using a stand-alone system just for managing maintenance inventory, which is not recommended as it is generally more expensive. Although the system can be developed to cover all specific maintenance inventory management needs, having an entirely separate system would require more administrative overhead. Plus, having too many systems for the workers to use is not necessarily a good approach— there are too many user names and passwords to remember.

An inventory system is useless if the data in the system are not accurate. Therefore, procedures and policies must be in place to promote the discipline in updating the system as events happen in real time. A periodic audit must also be performed to ensure the system data match the actual inventory. One way to ensure data accuracy is to have dedicated employees manage the inventory and conduct cycle counts. Cycle counting is an inventory control method by which storeroom personnel physically count a percentage of the total inventory items each day and correct the errors [12]. If resources are not available for dedicated inventory management, maintenance managers may consider limiting the inventory access rights to senior personnel such as forepersons or supervisors and hold them accountable for inventory accuracy. This brings up the topic of inventory security. It is preferred that the maintenance inventory, especially class A and B inventory, is stored in a locked area with security access to prevent unauthorized withdrawals. A computerized card reader system is highly recommended as it time stamps each access by the employees and keeps these data in the database.

Making sure the maintenance inventory is managed efficiently and accurately requires a great deal of discipline and ongoing real-time efforts. Many companies are outsourcing the inventory management functions to firms that specialize in logistics and inventory management. These firms provide an entire inventory management package that includes inventory management software, on-site personnel to handle receiving and issuing parts, cycle counts and regular audits of the inventory, ordering and delivering, and so on. It is an option for those companies that have high inventory volume and activities.

6

Maintenance Performance Indicators

INTRODUCTION

Indicators are essential to any business as they reveal the performance and status of the operations. Correct use of indicators can initiate the right actions needed to improve the business, and indicators serve as the means to measure the success of such actions. Maintenance is no exception. Terry Wireman, in his book *Developing Performance Indicators for Managing Maintenance*, stated that the objective of performance indicators is "to provide an overall perspective on the company's goals, business strategies, and specific objectives" [11, p. xvi]. He presented the maintenance indicators in the categories of corporate indicators, financial performance indicators, efficiency and effectiveness performance indicators, tactical performance indicators, and functional performance indicators. Over 100 indicators were listed in Wireman's book, and every aspect of maintenance management can have indicators to reflect its status.

However, indicators are the reflections of current and past information, which are like the rearview mirrors of a car. One who drives the car needs to check them occasionally but must not lose sight of where the car is heading. As such, maintenance managers should not put too much emphasis on the indicators and lose focus on what is coming up. Indicators are useless if a clear direction for the operations is absent. Therefore, indicators are necessary but should not be overdone as generating indicators requires administrative effort in data entry, data correction, data maintenance, as well as reporting. Many managers use indicators merely as nice charts that decorate the office or the hallway, creating an impression that they are the bosses who are closely managing the operations or even that they are bragging how well the organizations are managed. This is a misuse of indicators and a waste of time and resources. Indicators must have a higher purpose: exposing issues, driving for improvements, and measuring progress. If an area has been operating at near-optimal performance without any issue or customer complaint, regular indicator reports may not be needed. For instance, a machine is operating at 99% availability, which is beyond its 95% goal, and achieving 100% availability is statistically unfeasible. Generating extensive indicator sets to monitor the performance of this machine should not be necessary. Maintenance managers should direct the time and resources to other areas that need improvement and only check one or two key indicators of this machine occasionally.

This chapter discusses the key indicators that are used frequently in managing maintenance and classifies them into three intuitive categories: equipment performance indicators, process performance indicators, and cost performance indicators. Each indicator is presented with emphasis on its purposes in addition to its definition and calculation. Equipment performance indicators are equipment specific and for revealing equipment issues. Process performance indicators are related to the functioning of the maintenance operations and for reflecting the process issues in maintenance management. Cost performance indicators are associated with the financial management of maintenance operations and for controlling maintenance spending.

EQUIPMENT PERFORMANCE INDICATORS

Equipment performance indicators are the measurements of equipment status or historical trends. The key areas that measure equipment performance are safety, availability, reliability, maintainability, and utilization.

Safety Indicators

Purpose

Safety indicators are necessary for addressing equipment safety issues and reducing potential safety incidents, such as personal injuries and environmental damage caused by equipment failures or design flaws.

Format and Variation

1. *Number of potential safety incidents associated with a particular type of equipment.* The potential safety incidents are the least serious among all safety problems when safety issues are noticed before

personal injuries and environmental damages occur. Different companies name these differently, such as safety bulletin incidents (SBIs), safety near misses (SNM), or safety FYI (SFYI).

2. *Number of environmental incidents associated with a type of equipment.* Environmental incidents occur when equipment failures cause contamination of the environment, which could cause safety and health issues for the workers and the community. These could also lead to government agency investigations and fines.

3. *Number of first-aid cases associated with a type of equipment.* First-aid cases are the least-serious personal injuries; employees are medically treated but can return to work.

4. *Number of lost-day cases associated with a type of equipment.* Lost-day cases are a more serious type of safety incident; the injured employee must take days off to recover.

5. *Number of fatality cases associated with a type of equipment.* Fatality cases refer to personal death cases, which are the most serious safety incidents. It is hoped that such incidents will not occur as maintenance managers review the safety indicators regularly and address them before the issues become more serious.

Presentation

These safety indicators are typically presented in the bar chart format with a monthly or quarterly rolling scale, preferably starting when the equipment is first introduced. All the indicators can be included in a single chart with different colors representing the various safety incidents.

Availability Indicators

Purpose

The purpose of availability indicators is to monitor overall equipment performance. They are used for capacity planning to determine the number of machines needed for operations. If the availability is high above the desired goal, other equipment indicators such as MTBF (mean time between failures) and MTTR (mean time to repair) may not need to be generated and reviewed.

Note: Please review the section on key equipment terminology in Chapter 1 for the time definitions used in the calculations under this section.

Format and Variation

1. *Total availability:* This is a measure of equipment availability that accounts for the entire calendar time and all equipment states. Weekly calculations are recommended but can also be in the 4-week format.

$$\text{Weekly Total Availability} = \frac{\text{Equipment uptime}}{168}\%$$

$$4 - \text{Weekly Total Availability} = \frac{\text{Equipment uptime in 4 weeks}}{672}\%$$

2. *Operational availability:* This equipment availability indicator reflects operations time only.

$$\text{Weekly Operational Availability} = \frac{\text{Equipment uptime}}{\text{Operations time}}\%$$

where Operations time = 168 – Nonscheduled time.

3. *Manufacturing availability:* Equipment availability reflects manufacturing time only.

$$\text{Weekly Manufacturing Availability} = \frac{\text{Equipment uptime}}{\text{Manufacturing time}}\%$$

where Manufacturing time = Operations time – Engineering time.

4. *Equipment-dependent availability:* Equipment availability relates to the failures of the equipment only. It discounts all the equipment unavailability due to facility or environmental issues, network or information technology (IT) outages, human errors, and so on. Its intent is to measure the real performance of the equipment itself. It can be in the weekly, 4-week, and 13-week rolling average formats.

$$\text{Weekly Dependent Availability} = \frac{\text{Equipment uptime}}{\text{Equipment time}}\%$$

where Equipment time = Operations time information technology – All downtime unrelated to the equipment.

5. *Supplier-dependent availability:* Equipment availability relates to the failures of the equipment owned by the vendor only. It is a subset of equipment-dependent availability discounting all internal maintenance delays that have an impact on the equipment unavailability. Its intent is to measure the vendor support performance for contract and warranty management. It can be in the same format as the previous indicator.

$$\text{Supplier Dependent Availability} = \frac{\text{Equipment uptime}}{\text{Supplier equipment time}}\ \%$$

where Supplier equipment time = Equipment – All downtime-related internal maintenance delays.

6. *Availability breakdowns:* This indicator shows the availability categories as well as downtime that cause the availability losses. Downtime percentage is easily obtained from the availability numbers by using the formula

$$\text{Downtime } \% = 1 - \text{Availability}$$

When equipment availability indicators are below the goal, further breakdown of the downtime percentage is needed to reveal and address the main issues. This indicator helps to identify problem areas for the entire factory process, which includes operations and equipment management groups regarding who owns which areas of improvements.

Presentation

These availability indicators 1 to 5 are typically presented in the line chart format on a time scale. Multiple availability indicators can be presented in the same chart to reveal the gaps between them. For analyzing the trend, rolling 4-week and 13-week averages can also be used. In this case, it is typical to have weekly, 4-week rolling averages, and 13-week rolling averages in the same chart. With operations that have multiple identical machines, a fleet availability indicator can be calculated by averaging all individual equipment availabilities. It is also typical to have all individual machine availabilities and the fleet availability in the same chart to reveal performance variation among the equipment in the fleet. In addition, a goal or target line is typically included in all indicator charts for reference. For the availability breakdown indicator, a bar chart (with all the bars set at 100%) based on the percentages of the categories over time is often used. For a specific period, a pie chart illustrating all the equipment availability categories and different downtime percentages can be used as well.

Reliability Indicators

Purpose

These indicators provide statistical averages that measure reliability—how often the equipment fails. This indicator is used to drive failure frequency reduction, predict equipment failure occurrence, and forecast resources (i.e., headcount/parts). If the availability is high above the desired goal, frequent review of the reliability indicators may not be needed.

Format and Variation

1. *Mean time between failures.* (*Note:* This reliability indicator is a mean average, so a weekly number is not recommended. It is typical to use 4-week and 13-week calculations.)

$$4 - \text{Week MTBF} = \frac{\text{Equipment uptime in 4 weeks}}{\text{\# of failures in 4 weeks}}$$

When # of failure = 0, set it to 1.

$$13 - \text{Week MTBF} = \frac{\text{Equipment uptime in 13 weeks}}{\text{\# of failures in 13 weeks}}$$

When # of failures = 0, set it to 1.

2. *Productive MTBF.* This measures true MTBF that has an impact on production. It discounts all the engineering time and standby time.

$$4 - \text{Week MTBF} = \frac{\text{Productive time in 4 weeks}}{\text{\# of failures within 4 weeks' Productive time}}$$

When # of failures = 0, set it to 1.

3. *Mean units between failures (MUBF).* It measures equipment failure rate as a function of the number of production units. It reveals the correlation between failure rate and machine usage, rather than a time-based correlation. If a huge quantity of product is processed, this indicator can be based on weekly calculation; otherwise, use a 4-week or 13-week format. For your information, in the semiconductor industry, mean wafer between failures (MWBF) is an often-used indicator.

$$\text{Weekly MUBF} = \frac{\text{Total \# of production units processed}}{\text{\# of failures in the week}}$$

When # of failures = 0, set it to 1.

4. *Mean cycles between failures (MCBF).* It measures equipment failure rate as a function of the number of equipment cycles. Similar to the MUBF, it indicates the failure rate in relation to the rate of machines exercised.

$$\text{Weekly MCBF} = \frac{\text{Total \# of equipment cycles}}{\text{\# of failures in the week}}$$

When # of failures = 0, set it to 1.

5. *Mean time between assists (MTBA).* It measures the rate of interruption during operations. This indicator demonstrates the user friendliness of the equipment rather than hard failures. It is generally a focus of production managers. It is typically in 4-week and 13-week formats like MTBF.

$$4 - \text{Weekly MTBA} = \frac{\text{Productive time in 4 weeks}}{\text{\# of assists in 4 weeks}}$$

When # of assists = 0, set it to 1.

Presentation

The reliability indicators are generally presented in the line chart format on a time scale. Multiple reliability indicators can be presented in the same chart as well. For operations that have multiple identical machines, a fleet reliability indicator *cannot* be calculated by averaging all individual equipment reliability indicators, as in the case of availability indicators. Fleet reliability indicators are calculated as follows:

$$\text{Fleet MTBF} = \frac{\text{Sum of equipment uptime of all machines}}{\text{\# of failures of all machines}}$$

When # of failures = 0, set it to 1.

$$\text{Fleet MUBF} = \frac{\text{Sum of units processed through all machines}}{\text{\# of failures of all machines}}$$

When # of failures = 0, set it to 1.

$$\text{Fleet MCBF} = \frac{\text{Sum of equipment cycles of all machines}}{\text{\# of failures of all machines}}$$

When # of failures = 0, set it to 1.

$$\text{Fleet MTBA} = \frac{\text{Sum of productive time of all machines}}{\text{\# of assists of all machines}}$$

When # of assists = 0, set it to 1.

To simplify the calculations on the MTBF, sometimes the following formula is used:

$$4 - \text{Week MTBF} = \frac{4 \times 168}{\text{\# of failures in 4 weeks}}$$

When # of failure = 0, set it to 1.

This reveals the general trend of the failure rate under total time. For more accurate calculation, the following formula includes downtime, which is easily obtained and needed for the calculations of MTTR as well.

$$4 - \text{Week MTBF} = \frac{4 \times 168 - \text{Downtime}}{\text{\# of failures in 4 weeks}}$$

When # of failures = 0, set it to 1.

This calculation is generally accurate if the operation is running 24 hours a day and 7 days a week.

Maintainability Indicators

Purpose

These indicators provide statistical averages that measure maintainability: how quickly the equipment can be repaired or assisted. These indicators are used to understand equipment interrupt occurrences as well as to provide a measurement for maintenance skills for driving downtime reduction. The indicators are also used to forecast resources (i.e., headcount). If the availability is high above the desired goal, frequent review of the maintainability indicators may not be needed.

Format and Variation

1. *Mean time to repair.* This indicator should capture the repairs on equipment-related failures only. Other troubleshooting actions should be captured in other indicators, such as mean time to assist (MTTA) and mean time off line (MTOL), which will be discussed next. (*Note:* Similar to reliability indicators, a weekly number is not recommended. It is typical to use 4-week and 13-week calculations.)

$$4 - \text{Week MTTR} = \frac{\text{Total repair time in 4 weeks}}{\text{\# of failures in 4 weeks}}$$

When # of failures = 0, set it to 1.

$$13 - \text{Week MTTR} = \frac{\text{Total repair time in 13 weeks}}{\text{\# of failures in 13 weeks}}$$

When # of failures = 0, set it to 1.

2. *Mean time to assist.* This indicator is focused on the average time to assist the equipment when interruptions occur. It serves a similar purpose as MTBA with a production focus.

$$4 - \text{Week MTTA} = \frac{\text{Total assist time in 4 weeks}}{\text{\# of assists in 4 weeks}}$$

When # of assists = 0, set it to *1*.

3. *Mean time off line.* This indicator provides the average time in maintaining the equipment by including all down events of the equipment.

$$4 - \text{Week MTOL} = \frac{\text{Total equipment downtime}}{\text{\# of down events}}$$

When # of down events = 0, set it to 1.

4. *Repair breakdowns.* This indicator shows the major categories that cause the downtime on the equipment. It helps to identify equipment-specific issues for directing equipment improvement activities, vendor management, training issues, and so on.

Presentation

The maintainability indicators 1–3 are typically presented in the line chart format on a time scale. Multiple maintainability indicators can be presented in the same chart. Also, for the same equipment, MTBF and MTTR can be included in the same chart using the two-axis option on the *y*-axis for different scaling for optimal data presentation. Similar to reliability indicators, averaging all individual equipment MTXX indicators for the fleet indicator will not work. Fleet maintainability indicators are calculated as follows:

$$\text{Fleet MTTR} = \frac{\text{Sum of total repair time of all machines}}{\text{\# of failures of all machines}}$$

When # of failures = 0, set it to 1.

$$\text{Fleet MTBA} = \frac{\text{Sum of total assist time of all machines}}{\text{\# of assists of all machines}}$$

When # of assists = 0, set it to 1.

$$\text{Fleet MTOL} = \frac{\text{Sum of total downtime of all machines}}{\text{\# of down events of all machines}}$$

When # of down events = 0, set it to 1.

For the repair breakdown indicator, a bar chart based on the repair categories over time can be used. For a specific period, a pie chart illustrating all the equipment repair categories can also serve the purpose of highlighting the key repair areas that need improvement.

Utilization Indicators

Purpose

Utilization indicators measure the usage of the equipment, which relates to the bottom line regarding why the equipment is acquired. The utilization indicators are considered operations/production indicators and were seldom used by maintenance personnel under the maintenance era. The use and focus of utilization indicators for equipment management is a significant shift in the post-maintenance era that is discussed in later chapters. The utilization indicators are presented in this section so readers can easily find them in the indicator chapter. Under the maintenance setup, utilization indicators are used by maintenance personnel for reference only and belong to the nice-to-know category. Maintenance personnel could use these indicators to set priorities on resolving equipment issues when resources are limited. Highly utilized machines will be repaired first instead of machines with low utilization. The maintenance department does not own the generation of these indicators, and they are generally generated by the production or operations department for driving output. Most computerized maintenance management systems (CMMSs) do not have the capability to generate and present these indicators.

Format and Variation

1. *Total utilization.* Equipment utilization that accounts for the entire calendar time. Similar to availability indicators, weekly calculations are recommended but can also be in the 4-week format.

$$\text{Weekly Total Utilization} = \frac{\text{Productive time}}{168} \%$$

$$4 - \text{Weekly Total Utilization} = \frac{\text{Productive time in 4 weeks}}{672} \%$$

2. *Operational utilization.* This equipment utilization reflects operations time only.

$$\text{Weekly Operational Utilization} = \frac{\text{Productive time}}{\text{Operations time}} \%$$

where Operations time = 168 – Nonscheduled time.

3. *Period-specific utilization.* For driving improvement in equipment utilization, it is often necessary to generate indicators to measure the utilization for a specific time period. The examples would be prime-time, off-hour, weekend, and shift-specific utilizations. To generate these utilization indicators, the company must define the period. Using prime time as an example, some companies define this as Monday through Friday, 8 a.m. to 5 p.m.; others define this as Monday through Friday, 9 a.m. to 6 p.m. The formula would be as follows:

$$\text{Prime Time Utilization} = \frac{\text{Productive time within prime time}}{\text{Operations time within prime time}} \%$$

For operations that use equipment for engineering during prime time and production during off-hours, the prime-time utilization can be called the engineering utilization. Similarly, the off-hour, utilization, and shift utilizations can be calculated once the time periods are defined.

Presentation

Similar to availability indicators, the utilization indicators are typically presented in the line chart format. In fact, since they are all based on percentages, utilization and availability indicators for the same machine are generally done in the same chart to reveal the gaps between them. Multiple period-specific utilization indicators can also be presented together as well. Rolling 4-week and 13-week averages can also be used for analyzing the trend. The fleet utilization indicator can also be calculated by averaging all individual equipment utilizations, just like availability indicators. It is also typical to have all individual utilizations and the fleet utilization in the same chart to reveal equipment usage variation among the equipment in the fleet.

PROCESS PERFORMANCE INDICATORS

Process performance indicators measure how well the maintenance operations perform. There are many processing issues that could have an impact on equipment performance. Developing indicators to measure the process efficiency is needed to identify and solve these processing issues. In general, if the equipment performance is far above the required goal and management is satisfied with the current situation, time and effort in tracking the process performance indicators may be minimal. For companies that constantly focus on continuous improvement and drive for further budget and headcount reductions, these indicators will definitely need to be tracked and reviewed regularly. The key process performance indicators are labor productivity, nonproductive downtime, customer satisfaction, and operational misses.

Labor Productivity Indicators

Purpose

These indicators provide measurements of worker productivity in the maintenance operations.

Format and Variation

1. *Headcount/equipment ratio.* This indicator provides an overall trend for the support headcount versus changes in the equipment count. This ratio is also useful for planning when headcount estimates are needed for an equipment count increase or decrease.

$$\text{Headcount/equipment} = \frac{\text{Total equipment count}}{\text{Total headcount}}$$

In a typical factory, not all machines have the same complexity and service level. Some machines are big and cost millions of dollars, while others are portable and inexpensive. Therefore, to be effective in measuring the productivity of the maintenance organization, a variation of this indicator can be derived as follows:

$$\text{Headcount/equipment} = \frac{\text{Major equipment count}}{\text{Direct support headcount}}$$

This formula only counts major equipment as defined by management and the direct support headcount, which is the count of employees who physically work on the equipment. It does not include management, planning, and administrative staff. Management may measure productivity of the entire department by substituting total headcount in the denominator. Also, if indirect headcount productivity is needed, the formula can be changed to have indirect support headcount as the denominator.

2. *Work order/headcount ratio.* This indicator provides a measurement of the amount of work activities per worker.

$$\text{Work order/headcount} = \frac{\text{Total work order count}}{\text{Direct support headcount}}$$

3. *Downtime/headcount ratio.* This indicator is similar to the work order/headcount ratio but measures the hours rather than the number of incidents.

$$\text{Downtime/headcount} = \frac{\text{Total downtime}}{\text{Direct support headcount}}$$

4. Work order per shift and downtime per shift: These indicators provide the labor productivity among different shifts to drive for improvements if particular shifts are constantly having lower performance than the others.

$$\text{Work order/shift} = \frac{\text{\# of work orders worked by the given shift}}{\text{Headcount of the shift}}$$

$$\text{Downtime/shift} = \frac{\text{Total downtime in the given shift}}{\text{Headcount of the shift}}$$

Presentation

The labor productivity indicators are typically presented in the line chart format with a rolling time scale. Please keep in mind that these indicators are intended to reveal the trend, so management should not focus or overreact to the change of an individual occurrence or period. For high-level indicators such as headcount/equipment ratio, a quarterly scale is generally used. The weekly scale is fine for the other indicators. The labor productivity indicators listed are the popular ones, and there are others that can be used to address specific problems. For instance, work order per individual and repair time per individual measurements can be used if certain individuals have performance issues. However, these individual-focused indicators must be used sensitively and confidentially as their use may lead to an undesirable work environment that discourages teamwork and cooperation among workers.

Nonproductive Downtime Indicators

Purpose

The nonproductive downtime indicators highlight and provide trends for major non-valued-added wait times and events during scheduled and unscheduled downtime for addressing issues that lead to maintenance delays.

Format and Variation

1. *Response time.* This indicator provides a measurement of how quickly the maintenance group responds to work requests. This is typically related to unscheduled events only. Including response time to scheduled events such as upgrades and PM (preventive maintenance) will dilute the data and make this indicator less effective.
2. *Waiting for part time.* This indicator provides a measurement of the effectiveness of the spare inventory program.
3. *Waiting for support/personnel.* This is a group of indicator that includes wait time for different support groups and personnel for driving improvement with each specific group to reduce unproductive wait time. The typical nonproductive downtime indicators are waiting for maintenance tech time, waiting for equipment engineering time, waiting for management decision time, waiting for vendor time, waiting for facilities time, waiting for network/IT time, waiting for operations time, waiting for instruction time, and waiting for verification time.

Presentation

The nonproductive downtime indicators are generally presented in the stacking bar chart format on a weekly or monthly scale. If there is not a specific maintenance delay that needs attention, all nonproductive downtimes can be presented in the same graph. Another way to present the nonproductive downtime data is to plot all the wait times within a time period from high to low, which helps to identify the top issues. When there is a problematic area that requires management focus, the indicator of this area needs to be in its own chart to see the trend. Typically, response time, waiting for part time, and waiting for vendor time are plotted individually since they account for most of the nonproductive downtime in many companies and industries. Again, the indicators listed here are typical, and some operations have special processes and support groups, so custom wait time indicators will need to be generated to drive the solutions if there are issues.

Customer Satisfaction Indicators

Purpose

Customer satisfaction is measured as if a maintenance organization is a service organization serving internal customers.

Format and Variation

1. *Regular customer survey ratings.* Maintenance managers may send out questionnaires to key customers regularly to get feedback on the service and support provided to the customers. Typically, regular surveys are conducted quarterly, semiannually, or annually. The survey should cover customer responses, time to resolutions, communication, support personnel knowledge, and professionalism. To be effective, it should not have more than 10 questions.

2. *Real-time feedback after each work order.* It has been popular for service organizations, such as IT support groups, to set up computerized systems to send out a short survey to customers automatically on completion of a service. Many maintenance organizations are doing the same with their CMMSs. These short surveys typically have five questions or fewer and allow the customers to provide feedback quickly on the services while the support events are still fresh in their minds. One drawback of the real-time survey is that it captures the satisfaction of the equipment users only. Those who manage the equipment users, such as managers and supervisors of operations and product engineering, never physically submit a work order, so the survey would never be sent to them. They may have a different perception of the maintenance services. Therefore, it is typical that regular surveys are sent to management to supplement the real-time surveys.

3. *Customer request/complaint ratio.* This indicator measures the number of customer complaints in relation to the number of customer requests. This indicator generally covers the entire maintenance department, but it can be shift based to measure customer satisfaction with a particular shift if issues are suspected in the given shift.

$$\text{Customer request/complaint} = \frac{\text{\# of customer requests}}{\text{\# of customer complaints}}$$

Presentation

Customer satisfaction indicators 1 and 2 are presented as the percentage of favorable ratings or average scores of each question in the customer survey. The bar chart format is normally used with several time periods plotted together to indicate the trend and progress. Customer request/complaint ratios are typically plotted in line chart format over a time scale.

Operational Misses and Error Rates

Purpose

Measure the number of misses and errors in maintenance processes and events. Essentially, each maintenance process and logistic area can have its own operational misses and error rate indicators, which can be a significant time requirement for generating all these indicators. It is recommended that only the indicators for the suspected problematic processes and events be generated. The typical indicators are SWAT (from the military for special weapons and tactics)/escalations, spare outage, PM misses, and CMMS data entry errors.

Format and Variation

1. *SWAT/escalation.* Measuring the frequency of SWAT and escalation occurences can provide an understanding of the difficulty factor in maintaining the equipment and the internal skill levels. SWAT is an equipment hard-down situation for which a task force is required to identify the root cause and solve the issues (SWAT definition can be found in Chapter 5). Escalation is when the internal maintenance personnel can no longer perform the repairs and need to involve management and vendor support. This indicator can be just in the number of occurrences over time or as a rate by dividing the total number of equipment down incidents.

$$\text{SWAT rate} = \frac{\# \text{ of SWATs}}{\text{Total} \# \text{ of down incidents}}$$

$$\text{Escalation rate} = \frac{\# \text{ of escalations}}{\text{Total} \# \text{ of down incidents}}$$

2. *Spare outage.* This is an indicator for the spare inventory program as it measures the frequency of spare part outages. It can also be in the number of occurrences over time or as a rate by dividing the total number of replacement parts requested.

$$\text{Spare outage rate} = \frac{\text{\# of spare out of stock}}{\text{Total \# of spare requested}}$$

3. *PM misses.* This indicator measures the discipline of the maintenance department in carrying out the PM. It can be just the number of PM misses in a given time period or as a rate of PM misses over the total number of PM performed.

$$\text{PM miss rate} = \frac{\text{\# of PM misses}}{\text{Total \# PM performed}}$$

4. *CMMS data entry error.* This indicator measures the accuracy of the data entry into the CMMS as the accuracy of maintenance performance indicators is based on the correctness of the data entry. This indicator can also be just the number of errors in a given time period or as an error rate over total number of records in the CMMS.

$$\text{CMMS data entry error rate} = \frac{\text{\# of data entry errors}}{\text{Total \# records}}$$

This indicator can go into further detail to measure error rate of the different shifts and individuals to drive data accuracy if in fact it is a big problem.

Presentation

Operational misses and error rates indicators can be presented as the number of occurrences over a time scale or as a rate of the total events over time. Both line chart and bar chart formats are used. For the indicators presented as rates, sometimes they are plotted in a bar chart with percentages shown. The indicators illustrated are the frequently used ones, and many other indicators can be generated for different operation models and situations.

COST PERFORMANCE INDICATORS

Already mentioned in the budgeting section, maintenance is an overhead function, so there is enormous pressure for cost saving. Another characteristic of maintenance spending is that the operation is dealing with uncertainty so the spending range from month to month or quarter to quarter is huge. Many finance personnel are trained to manage the budget of an operation based on run rates and spending trends, which is a typical tactic learned from classes in a master's of business administration program. Since maintenance spending often fluctuates and has no appearance patterns or trends, corporate finance personnel often consider maintenance spending out of control. Therefore, maintenance managers must pay extra attention to provide detailed cost performance indicators to justify the uncommon spending reality of maintenance operations.

Cost Rates

Purpose

The purpose of this indicator is to provide a general understanding of the cost spending levels and trends for the maintenance activities. Also, cost behavior changes are tracked to alert maintenance managers to take proper actions and to allow quick estimations for factory capacity and maintenance activity-planning purposes.

Format and Variation

1. *Cost per equipment.* This is a high-level cost indicator that measures the overall cost performance based on the entire equipment base. It can also help the factory to determine quickly the cost for additional equipment increase in capacity planning and in the what-if analysis.

$$\text{Cost per equipment} = \frac{\text{Total maintenance cost of equipment}}{\text{\# of equipment}}$$

Since the equipment base is typically made up of different machines ranging from simple to complex, this indicator can be further broken down to specific equipment groups and types, such as cost per generator, cost per pump, cost per tester, and so on. In these cases, the sublevel indicators can be obtained by substituting the word *equipment* with the specific machine type in the formula just given.

2. *Cost per (activity)*. Activity is substituted for the specific maintenance events, so the indicators in this category are cost per repair, cost per PM, and cost per setup.

$$\text{Cost per repair} = \frac{\text{Total repair cost}}{\text{\# of repairs}}$$

$$\text{Cost per PM} = \frac{\text{Total PM cost}}{\text{\# of PM}}$$

$$\text{Cost per setup} = \frac{\text{Total setup cost}}{\text{\# of setups}}$$

Cost per upgrade is used sometimes, but this indicator will not have much practical value unless the upgrades are done frequently and the work scopes of the upgrades are relatively consistent. Knowing the historical averages and trends of these cost rates on the key maintenance activities enables maintenance managers and planners to assess the budget needed quickly based on the anticipated maintenance activity levels.

Presentation

The cost rate indicators are typically presented in the line chart format over a time scale. Multiple rates can be presented in the same chart for easy viewing and comparison. For analyzing the trend, rolling 4-week and 13-week averages can also be used for individual cost rate.

Cost Breakdown by Categories

Purpose

Detailed level cost breakdowns are used to justify and track spending in different maintenance activity categories. This provides a better understanding of the spending behaviors of each area and enhances the accuracy for future budget forecasts.

Format and Variation

1. *Fixed and variable costs.* Breaking down maintenance costs into fixed and variable costs and then presenting the two categories of spending over a time scale helps to separate the uncertainty factor in maintenance spending. Fixed cost should have a clear run rate pattern as typical operations with minimal uncertainty. Any increase or decrease in spending can be linked to a justifiable reason. Variable cost typically does not have a clear pattern but can be correlated to equipment performance indicators discussed in previous sections. Notes are typically provided on the peaks and lows in the chart to explain the changes in spending.

2. *Repair and PM costs.* Dividing maintenance costs into repair and PM is another way of isolating the uncertainty factors in maintenance spending. PM spending should have a more stable pattern than repair spending. These indicators are less used compared to the breakdown of fixed and variable costs as they are not applicable when equipment PMs and repairs are covered under contracts with fixed cost.

3. *Direct and indirect costs.* Separating maintenance costs into direct and indirect helps to understand the efficiency of the maintenance operations and guides maintenance managers to spend money in the right places. Direct costs are associated with equipment maintenance events such as repairs, PM, upgrades, and setups. They can be further broken down to labor and material costs. Indirect costs are overhead costs that can be broken down to tool cost, training cost, CMMS cost, spare management cost, administrative cost, and so on. The detailed breakdowns allow maintenance managers to control the spending in each area.

Presentation

The combined cost breakdown indicators are typically presented in the stacking bar chart format over a time scale. For individual cost presentations, such as just presenting the fixed cost, the line chart format over a time scale is generally used as it is better for demonstrating the trend. Detailed breakdown of direct and indirect costs may be done in the pie chart format for a specific period. This does not show the trend but reveals the big spending areas that need further attention.

7

Computerized Maintenance Management Systems

INTRODUCTION

The application of computers for productivity improvements is common in most operations due to the explosive growth of computer and software availabilities and capabilities. Maintenance is no exception. The widespread use of computerized systems for maintenance operations started in the mid-1980s, with a significant growth period in the 1990s. The term CMMS became one of the most popular terms in the maintenance vocabularies in the 1990s. CMMS stands for computerized maintenance management systems by most people or computer-managed maintenance systems by some. In either case, it is defined as the computer-based solution to improve maintenance efficiency. In the 1999 book *Computer-Managed Maintenance Systems* by Cato and Mobley, it stated that there were over 300 vendors who offered CMMS solutions, with prices ranging from less than $1,000 to over $1 million for implementation at a single site [8].

The development of the CMMS started out as simple spreadsheet and database programs for keeping track of maintenance records, moving to today's enterprise solutions with local-area network (LAN) and Internet/intranet access connecting all responsible parties in maintenance and the equipment. They cover every aspect of maintenance management, including the enabling of predictive maintenance (PdM). Despite ample commercially available CMMS packages, many companies use self-developed CMMS solutions. The typical reasons for developing in-house solutions are cost saving and the required customization for the company's operations. Purchased or developed, many companies spend a significant amount of time, money, and resources to commission CMMSs. However, not all companies are successful in getting the desired benefits from the CMMS. In other words, many CMMS implementations fail. There are numerous reasons why a CMMS fails, which can be related to system technical issues, managerial and planning issues, irrational expectations, poor communication and training, and mismatching corporate or work group cultures. Maintenance managers must remember that CMMS is just a tool to help manage the maintenance operation. Do not use CMMS just because others are using it. If the use of computers leads to overwhelming administrative tasks and becomes a burden for everyone, then the return on investment is minimal as it hinders productivity instead of improving it. Putting such effort directly into equipment performance improvement projects may be a better use of resources. Also, different situations require different tools, so it is not one size fits all. Setting the right objectives for the CMMS is the first step for a successful implementation of an effective CMMS.

CMMS OBJECTIVES

Implementing a CMMS should be based on the characteristics of the equipment base and the operating model of the maintenance department. a CMMS may not be suitable for some equipment and some operational environments. CMMS as a tool cannot help if the goal is not clear. Therefore, before implementing a CMMS, the maintenance manager should have a clear priority in the business objectives and then select the CMMS objectives that align with the business objectives. Maintenance managers should choose cost-effective CMMS solutions that can achieve the desired objectives rather than using the shotgun approach of implementing a comprehensive system that covers everything. For instance, if the equipment performance is already at an optimal level but customer communication is a problem, the maintenance manager can select a simple CMMS that provides effective customer notification of equipment status and activities, thus avoiding a costly and time-consuming implementation of an enterprise CMMS package. Common business objectives and their corresponding CMMS system-level objectives are demonstrated in Table 7.1.

- *Provide PdM capability.* With the help of a computer, continuous monitoring systems can be developed to enable PdM, which means performing the right maintenance actions at the right time, therefore reducing preventive maintenance (PM) time and increasing availability of the equipment.

TABLE 7.1

Common Business and CMMS Objectives

Business Objectives	Key CMMS Systems Objectives
Improve equipment performance: availability, reliability, and maintainability	Provide predictive maintenance capability
	Fast notification to reduce response time and waiting for support personnel time
	Effective spares management to reduce part outages and waiting for part time
	Provide indicators for problem identification and solution
Increase productivity, quality of services, and customer satisfaction	Provide accurate record keeping
	Effective work order management
	Effective equipment activity scheduling
	Provide knowledge database for developing equipment support skills
	Effective customer communication
Increase cost performance	Reduce administrative labor
	Reduce paperwork
	Reduce inventory storage
	Provide cost data for accurate budget plan

- *Fast notification.* Quick notification of equipment issues can reduce response time and therefore increase equipment availability by reducing total downtime. It can also reduce the waiting time for other equipment support-related personnel when the ownership of the equipment issue changes, such as the escalation to equipment engineer, vendor, information technology (IT), or facilities support personnel.

- *Effective spare management.* An effective spare inventory management system can ensure on-hand inventory and reduce the chance of spare part outage, hence reducing total downtime. It also allows quicker access to the parts and therefore reduces waiting for part time.

- *Provide indicator reports.* Providing equipment performance indicator reports is one of the main objectives of most CMMSs. The right equipment performance indicator report can help management identify the key equipment issues and to direct and prioritize the resources and actions in problem resolution tactics. Reducing top equipment issues will lead to reliability improvements. This system-level objective can also be under the other two business objectives as most CMMSs are also capable of providing process productivity and cost performance indicators to help improve maintenance productivity and reduce costs.

- *Provide accurate record keeping.* Keeping accurate maintenance records is often part of the business requirements, such as when working on government contracts or meeting the International Organization for Standardization (ISO) 9000 standards. Accurate equipment records can significantly reduce the effort in asset audits and in configuration management.

- *Effective work order management.* The main business of the maintenance department is performing work orders. An appropriate work order system ensures that all the necessary information is captured for proper actions and the work requests are routed to the workers with the correct skills. The system also serves as the communication tool between the equipment user and the maintenance personnel. Work request details, status, owners, and actions taken can all be seen in the system. A work order system that is embraced by the customers will definitely lead to higher productivity and customer satisfaction.

- *Effective equipment activity scheduling.* Computerized scheduling tools are commonly used in most operations to improve productivity. In maintenance, although many activities, such as repairs and assists, are typically unscheduled events, PM and upgrades can be better managed with a computerized equipment activity scheduling tool, especially when PM is not calendar based. It allows better planning of resources and better user communication as well.

- *Provide knowledge database.* CMMS can be used as a knowledge depository for accumulation of equipment failure causes and troubleshooting actions. Maintenance personnel can search the CMMS to find out previous failure causes and resolutions when encountering similar issues, especially for workers who are new to the equipment. It helps to bring up the skill level of maintenance personnel in the entire department.

- *Effective customer communication.* One of the top customer complaints to the maintenance department concerns lack of communication on the equipment status and work order status. A CMMS can serve as a central information hut for all equipment information so users can pull up real time or historical information on any piece of equipment. The system can also be set up to notify the users automatically of state changes via e-mail or text messaging. These system features significantly reduce communication issues between the maintenance department and the user community.

- *Reduce administrative labor and paperwork.* Although implementing a CMMS typically takes some administrative efforts at the beginning, over the long run the right CMMS should save the maintenance department administrative labor. If the CMMS requires a great deal of administrative effort to keep it running, either the system is not the right choice or the implementation process is poorly carried out, that is, system administrators and users are not properly trained.
- *Reduce paperwork.* Using a CMMS for record keeping results in paperwork reduction by converting paper records to electronic records, which is good for the environment by saving some trees and the resources used in papermaking. The company also saves money on paper and record keeping.
- *Reduce inventory storage.* A CMMS with good parts and materials management capabilities can correlate equipment failure data to maintenance inventory records to optimize the stocking level of on-hand parts and materials. It helps to identify unnecessary spares and materials and typically reduces inventory storage costs.
- *Provide cost data for accurate budget plan.* Using a CMMS to keep track of detailed spending provides the data to increase the accuracy of a future maintenance budget plan, which increases the effectiveness of cost performance management.

As mentioned in Chapter 3, equipment management is transitioning out of maintenance management. There is a trend that equipment becomes more complex, and the modular approach is often used. It is typical that a manufacturing task is done by several modules combined, and these modules are made by different vendors. In some cases, these modules must be moved to different stations to support variation of manufacturing and engineering purposes. To manage these new changes in equipment, traditional CMMSs will not meet the objectives if the company decides to migrate to the post-maintenance era. The computerized equipment management systems (CEMSs) will meet the new requirements. The details of a CEMS and comparison to a CMMS are illustrated in Chapter 9.

CMMS FUNCTIONS

CMMS has a wide range of capabilities, from enterprise packages to simple home-grown database programs. After the desired objectives are identified by management, the right CMMS solution and its capabilities can then be determined. Many CMMSs are developed based on the modular approach, especially on commercially available CMMS packages. The CMMS developers typically offer CMMSs with a set of basic modules and have optional modules available for additional costs. For companies that choose to develop the CMMS in-house, the modular approach is also recommended as it provides flexibility and fast implementation. The typical functional modules offered in popular CMMSs are discussed in this section.

Equipment Module

The equipment module serves as a central depository of equipment-related information. It tracks equipment data, such as serial numbers, date of manufacturing, date of commissioning, configurations, upgrade history, acquisition costs, installation costs, upgrade costs, purchase agreements, service contracts, user manuals, vendor contact information, and the like.

When implementing this module, special attention is needed on the equipment-naming convention. Each equipment identifier must be unique in representing a piece of equipment. Typical identifiers should be intuitive by containing codes that reflect the equipment location, type, make or model, configuration, and the sequence of acquisition. Some companies use more finance-centric naming logic to include asset accounting codes, owning cost center or department codes, and so on. The identifiers generally begin alphabetically and end numerically. The last few numbers either represent the equipment sequentially as installed in the factory or use the serial numbers or a portion of the serial numbers from the equipment manufacturers. Using clear and consistent naming logic can reduce operational errors and even speed up maintenance response.

Work Order Module

The work order module allows work order generation, assignment, notification, update, closure, retreat, and record keeping. It is the main module for most CMMSs, so its functionality is generally the key factor in selecting a CMMS. The design of the work order module must match the company's operational model and equipment characteristics. It is also the main reason for companies to develop their own CMMS instead of purchasing one.

Equipment users typically initiate the unscheduled work orders, but if predictive maintenance is an objective of the CMMS implementation, choose a CMMS that has PdM capability with the interface to the equipment, which allows automatic generation of work orders based on preset conditions. After a work order is initiated in the system, the next step is assigning the work order to the right support personnel. Some CMMSs route all work orders to the management or planning staff to physically allocate work orders to technicians on duty, while most CMMSs use predetermined information such as service-level agreement (SLA) and support structure to route the work order to the right owner. While the work order is opened in the system, most CMMSs allow real-time updating of the work order status as frequently as required. Maintenance support technicians and management staff can change equipment states, reallocate work order owners, input messages, and so on. Equipment users can look up the sequence of events and up-to-date status of the work orders. Work order closure is typically done by the support technicians, with a few CMMSs requiring management final sign-offs.

Preventive Maintenance Module

The PM module keeps track of the PM schedules, PM task checklists, and materials lists. It allows maintenance managers or planners to schedule different types of PM easily in advance. Time-based PM scheduling is common in most PM modules, with a few CMMSs offering the capability to set up other PM types, such as input and output unit based, machine cycle based, or even condition based. In addition to managing the PM schedules, the PM module helps to ensure the PM is carried out in a timely and correct manner. Therefore, the chosen CMMS should generate PM work orders automatically based on the schedule and provide early notifications to the maintenance personnel and equipment users. It should also alert the maintenance personnel on parts and materials required in advance to avoid waiting for parts or delaying PM. Furthermore, the PM module should keep the PM task checklists and calibration records. Many CMMSs have the capability to prompt and guide the PM activities step by step and allow task checkoff electronically on completion. It also time stamps the activities for record keeping and operational efficiency studies.

Safety Module

The safety module keeps track of safety records and documents, such as Material Safety Data Sheets (MSDSs), hazardous materials used by the equipment, safety buy-off acceptance documents and dates, lockout and shutdown procedures, safety incidents, permits, and so on. Some CMMSs can automatically print the up-to-date lockout procedures or shutdown procedures on the work orders based on the nature of the work requested.

Labor Module

A full-scale labor module can track all maintenance personnel profiles and personal records, such as skill levels, training records, certifications, pay rates, salaries, work schedules, attendance, vacation, sick days, and overtime. Most corporations already have their human resource (HR) system, so most CMMSs do not have the full-scale labor module except for the enterprise-level CMMS. Typical CMMSs have labor modules that complement the HR system and keep track of items that are not under the HR system, such as the skill proficiency on different equipment types or tasks as well as training records and certifications related to the equipment. For maintenance operations that assign owners and support groups to the specific type of equipment or tasks, the labor module must have a support personnel database and the capability of linking it to the equipment and work order module, so the system can route work requests to the right equipment owners or the support technicians who are certified to take on the particular tasks.

Inventory Module

The inventory module manages the maintenance inventory, such as spares, tools, and materials. It is similar to a library management system that keeps track of the categories and the location of inventory as well as manages the checkout and return of the inventory. The naming conversion of the spares and materials also requires as careful consideration as that for the equipment identifiers. The detailed requirements are presented in the inventory management section in Chapter 5. The inventory module should allow setting minimum and maximum levels, flagging warranty parts to avoid unnecessary purchases, referencing internal part number to vendor part numbers, and tracing order information. Some CMMSs have ABC analysis capability, which allows the prioritization of inventory items. Some employ just-in-time methodology by utilizing the usage and lead time data in the system to alert the suppliers to prepare orders. Some inventory management systems can even generate and send purchase orders automatically to the suppliers based on the triggers and restocking algorithms defined in the systems.

Financial Module

The financial module records and controls the costs spent in maintenance activities, such as contract costs, labor costs, and materials costs. Similar to the labor module complementing the corporate HR system, the financial module in CMMS complements the corporate accounting system by keeping detailed spending records associated with individual equipment type or sometimes even individual machines to enable specific vendor management. It should preserve the corporate general ledger codes so data can easily be uploaded to the corporate system. Some CMMSs allow user-defined financial calendars to correlate with the corporate accounting periods. Purchasing support is also seen in many CMMSs. While some CMMS packages have the purchase function under the financial module, others have a stand-alone purchasing module. They can generate purchase requisitions or purchase orders and route them through the management and finance chain for approvals and authorizations. They also keep invoice and quotation information as well as supplier certification and performance data.

Calendar Module

The calendar module is sometimes called the scheduling module. It keeps the schedule of equipment activities in a calendar view, from production, engineering uses, to PM, upgrades, and setups. It is basically an activity planner. Every day, the maintenance personnel can look at the calendar to find out who is scheduled to use the equipment, what activities are taking place, and what maintenance tasks are scheduled. While some CMMSs limit the activity scheduling to management staff, others allow the equipment users and the general public to book equipment time and submit timed requests. The CMMS that has open access on requests but also an approval feature is highly recommended because it reduces administrative workload for the maintenance managers and planners to do data entry but still keeps control of the equipment. Essentially, two separate calendar views are required, one for managers and planners showing the requests and the other as the official approved calendar for the general public, which can be posted on the equipment or shown electronically in the equipment control computer as an access screen or screen saver.

CMMS FEATURES

The functions of the CMMS are task oriented, which enables the particular tasks to be carried out. The features are process oriented, which allows the tasks to be done in an efficient and user-friendly manner. In addition to evaluating the functions, management must consider the features when selecting or developing a CMMS. The popular features are discussed next.

Accessibility and Security

First-generation CMMSs were stand-alone programs. Second-generation CMMSs were connected through a LAN and were operating system (OS) dependent. Those were in the time when the UNIX OS was typically used in equipment control and a personal computer (PC) disk operating system (DOS) was used for management, planning, and office staff. The CMMS designed in the equipment control computing environment had the benefit of a direct equipment interface, but full access through the PC environment was difficult. In reverse, a CMMS with programs written in PC environment had difficulty in interfacing with the equipment. Therefore, accessibility was generally confined within the manufacturing facility or in the offices. The computing OS environment has changed significantly since the late 1990s, with improvements of PC OSs and the introduction of new OSs, such as Linux. CMMS accessibility has improved significantly, and the current generation of CMMSs has Internet connectivity that can be accessed anytime, anywhere through mobile devices such as notebook computers, smart phones, and personal digital assistants (PDAs).

Access and security are closely tied together. Most CMMSs have a different level of access permissions based on individual user accounts. Typical CMMS access levels are demonstrated in Table 7.2. With increasing accessibility, security becomes a bigger concern. Encrypted data and tighter access control protocols are applied. Using a CMMS with strong password

TABLE 7.2

Typical CMMS Access Levels

CMMS Access Level	Access Area	Work Scope
Administrator level	Full access	Perform system improvement and maintenance
Manager level	Systemwide data read-write access, report generation, and approval authorization rights	Manage the maintenance operation through CMMS: Control data and manage all maintenance resources, including personnel
Planner level	Manager level minus access rights to personnel information and selected approval rights	Plan, coordinate, and control maintenance resources and activities
Equipment owner level	Equipment type-specific data read-write access and limited approval authorization rights	Plan, coordinate, and control activities around the assigned equipment
Maintenance tech level	Work-related data entry and update rights	Perform maintenance tasks following system instructions and update the system with findings, actions, and resolutions
Operator and user level	Work order request rights and read-only rights in selected work-related information	Flag equipment issues and check equipment status for best equipment utilization

enforcement rules and automatic password expiration after a defined period is recommended.

Communication and Notification

One of the key efficiency improvement targets in maintenance management is reducing nonproductive wait time. CMMSs play a significant role in helping to achieve wait time reduction by quickly informing the right owners for the right actions. CMMSs typically provide signaling, e-mailing, and text messaging capabilities. Signaling refers to setting off an audio alarm or flashing light when the equipment needs attention. If an audio alarm is used, make sure it would not be confused with fire and safety alarms. Using low-volume beeping or a musical tune is preferred. The visual signaling devices are typically set on top of the equipment so they are visible around the factory floor so the technicians can quickly identify the particular machine. Different color lights or different flashing frequencies can be used to represent different issues.

The e-mailing feature is straightforward. On state changes or human initiation, an e-mail can be sent to a group list or an individual with the data from the CMMS. This feature certainly increases the communication between the users and the maintenance department, but it should not be overdone. Too many e-mails, especially long e-mails with the same format from the same account, annoy people. After a while, people just set automatic rules to delete them without reading. Typical CMMSs have a system e-mail account so all the recipients know the e-mails are from the CMMS. This is actually not a preferred setup. It is recommended that the CMMS user e-mail account be used to give a personal touch to messages. For instance, when closing the work order, an e-mail is sent to the work order requester through the e-mail account of the maintenance technician who resolved the issue. The message format and length also deserve careful consideration. Maintenance managers should seek inputs from the CMMS users to optimize message content design. Text messaging is another feature of CMMS. It started in the days when texting pagers were popular. In recent years, it has been replaced by instant messaging (IM).

Data Entry and Presentation

As people often say about the data in computerized systems, "It is garbage in and garbage out." Ensuring data accuracy in the CMMS is not an easy task. Automating the CMMS data entry as much as possible can definitely help to increase data integrity. The common features used in data entry are (1) setting defaults on data fields, (2) pulling the information from the equipment configuration files, (3) using predefined pulldown menus and checklists, and (4) utilizing bar code reading, magnetic card reading, and smart card (also called integrated circuit card) reading technologies. Careful design of data fields and input options along with a rigorous user training program are strongly recommended. The users must be educated on the intention of each data entry field and the meaning of each available choice. Avoid using vague input options such as "miscellaneous" and "others" because these options offer easy choices for people who do not want to take the effort to comprehend the system. It is typical to find that these indistinguishable options are bigger hits on the Pareto charts but offer no value for management.

Putting good data into the system is important, and how the data are presented also matters. Presenting the right data in the right format can help management identify issues and track progress toward improvement programs. The CMMS must have the capability to generate reports for the indicators discussed in Chapter 6. Time-based reports and real-time reports are common in most CMMSs. Time-based reports are generated regularly, such as daily, weekly, monthly, and so on, to reveal what happened in the past. Real-time reports are used to capture the current status of the equipment, the operation, and the resources. The "airport" display feature is a prime example. For instance, a CMMS can display a summary of current work orders on a big-screen TV with color codes. Another use is displaying the equipment status instead of the work order status. In addition to making the reports (pull reports) available in the system for the users to access, the CMMS should have the capability to send reports (push reports) automatically to the designated users. Some CMMSs have a report subscribing feature that allows the users to select the reports that they want and the time and the frequency that they want to receive such reports.

The format of the reports also warrants careful consideration. Presenting the data in tables and charts is better than just in text format. Tailor the chart formats for the indicators as described in previous sections for each indicator. Use color intuitively and consistently to highlight areas that need attention. Putting goal and limit lines in the charts also enhances the data analysis ability. Selecting CMMSs that have the option of generating statistical charts with control limits is also preferred.

Integration

Integration is the ability of the CMMS to interconnect to the equipment and other computerized business management systems, such as accounting systems, HR systems, production work-in-progress (WIP) tracking systems, knowledge/expert systems, and even office software. The main benefit of integration is elimination of duplicate data handling, which not only saves time and resources but also speeds up communication. It also indirectly results in data consistency and accuracy. Other benefits of integration are increasing the ability to cross-reference data and helping to identify hidden issues in the business. As mentioned, integration used to be a difficult task for CMMSs due to different computing OS environments. It is a big selling point for many popular CMMS systems, such as MAXIMO [13,14]. With the advancement of Internet-friendly and platform-independent programming languages, integration is easily achieved even for the systems that do not come with the capability. As long as it has a Web interface, a custom interface to any desired system can be attained by a typical software engineer in a relatively short time. In fact, many companies have done it by hiring college interns. A couple of integration examples are discussed here.

First, many systems are integrated with equipment control and have the ability to interact and control the operations of equipment. For instance, if a machine is down, the system on which the event is logged could provide software interlocks on certain functions of the machine. The interlocks can prevent people from starting the machine by mistake, thus avoiding equipment damage, product damage, or even physical harm to the repair personnel or to the operator.

Second, integration with a WIP tracking system is also very common. It provides many benefits. If a product has failed during the final test, the process can be backtracked to determine where exactly the problem occurred and on which specific machine the product was processed. Suppose, for example, that some electronic parts containing small cracks were discovered, but they passed the tests electronically, and the problem was intermittent and not detected until the final quality check. The products might have undergone several processes in the factory, and each process might have many identical equipment cells. If it was unknown what equipment cell tested the chips, the entire factory might have to be shut down to inspect all the machines. This scenario would also apply to opposite cases in which products may have come into the factory with defects that caused damage to the machines at testing. Integration between equipment and product process tracking allows managers to identify product problems caused by equipment failures or equipment failures caused by product defects.

Flexibility and Customization

Different operations have different characteristics, and there is no such thing as one size fits all in business applications. Also, the business environment is dynamic, and changes are often required. Implementing a CMMS takes time and effort, so no one wants to see that a system is obsolete or needs major modification in a short period after implementation. Selecting a CMMS that has flexibility and customization features can help companies to react quickly to the business changes and maintain effectiveness in managing the operations. Many CMMSs offer custom apps and utilities to create user-defined data fields, tables, and reports. Some provide programming capabilities in popular languages or system-specific macros that allow the users to create custom modules. This feature is specifically important for operations that want to implement PdM since PdM typically requires custom programs that interact with various types of equipment.

CMMS IMPLEMENTATION

Frequently seen implementation issues are poor system design, lack of user acceptance, lack of adequate training, and poor data input and utilization. It is important to select a CMMS based on the current organizational operating model, culture, and environment. Most CMMS implementation failures are because the CMMS functions and features are not in alignment with the way people perform the tasks. Some managers may argue that implementing a new CMMS is the means to drive a significant cultural paradigm shift to improve organization performance. In that case, management should focus on implementing the organizational change by consulting with the HR group. Management and HR need to do enough prework on changing employee mindset and morale rather than just implementing the CMMS and seeing the cultures collide. Otherwise, the implementation typically ends with failure for both intentions, and the CMMS is a sacrifice. Management then blames the failed CMMS and uses it as an excuse for the inability to make the organizational change. The suggestion is not to mix CMMS implementation with organizational change. Successful CMMS implementation requires a clear focus.

The CMMS implementation process follows the standard software system development life cycle. First, it starts with project initiation and planning when the CMMS implementation is proposed and approved with the objectives clearly defined. These objectives were discussed previously in this chapter. The second phase is requirements analysis, which translates the objectives into functional requirements of the CMMS. The third phase is solution definition, which identifies the possible solutions in achieving the objectives. In this phase, management faces the choice of buying or self-developing the CMMS. The decision is based on the performance requirements, the available costs, and the internal software development resources. Despite the large number of CMMSs available commercially, most companies actually develop their own CMMS because its effectiveness depends on how closely it is tailored to the equipment characteristics and the company's business model. Also, it is typical that much customization work is still required with most purchased CMMSs. With the advancement of programming languages and the increasing availability of skilled software programmers, developing an in-house software solution is becoming a more attractive approach for many companies. Management should seriously consider the self-development option.

The fourth phase is design, build, and configuration. When the purchase option is selected, the next step is evaluating available CMMSs and selecting one that meets the operational needs. Although the CMMS programming portion is completed on purchase, configuring the CMMS based on the company's operating environment and equipment base is inevitable. Therefore, maintenance management must work closely with the vendor to ensure the total package is delivered and technical support is available for the entire implementation phase. If the in-house development option is chosen, the next step is identifying the software engineering resources to design and build the system. The development of large systems is often done by forming a software project team or contracting it to a software developing firm. A hybrid approach is also commonly seen as the design of the system is done internally and the coding part is contracted out. Small systems can be done by utilizing internal software engineers. In fact, many small-scale CMMSs are done by internal resources on a part-time basis as typical continuous improvement (CI) projects. Hiring temporary workers and interns is another common approach.

After the system is installed and configured by either the vendor team or internal resources, the fifth phase is testing, which is conducted in typical software testing process steps, such as alpha testing, beta testing, and final testing. Alpha testing is done within the software development team. Beta testing is conducted with limited users or in selected locations. Final testing extends the system verification to include the general public. System training is often carried out during this phase to ensure the system is used correctly. Based on the bugs discovered or feedback provided during the testing phases, revisions of the system are released in a controlled fashion.

The final phase is the full release of the system—system documentations and administrative functions transitioning to the internal support staff and reenforcement of the discipline in using the system properly and in a timely fashion. Regular audits and ongoing training may be needed to maintain data accuracy in the system. Management should keep a log for all the potential improvement items and evaluate the possibility of implementing them in the next version or the next new CMMS.

Section III

The Post-Maintenance Era

8

The Systems View of the Equipment Management Process

INTRODUCTION

As equipment management progresses further away from the maintenance era, maintenance has developed a negative connotation for many people. Without changing the general maintenance approaches, management reinitiates the focus by repackaging the maintenance practices into improvement programs and picking catchy names such as DEEP (downtime elimination and education program) [4] and EQUIP (equipment quality unified improvement program). However, it is best to develop a new approach that meets the needs of the new environment.

Since the development of the general systems theory by Von Bertalanffy, many studies have been conducted to understand systems. Systems thinking and a systems approach have become increasingly common in organizations. Many practical systems models have been developed to guide management to resolve complex problems at the organizational level. One of these models is the organization systems model developed by Kast and Rosenzweig.

Kast and Rosenzweig viewed an organization as an open system operating with five subsystems: goals and value, technical, psychosocial, structural, and managerial, all of which operate under the environmental suprasystem [15]. Although this model is developed to analyze an organization as a whole entity, it can also be used to analyze a division of an organization in which complex problems exist. As described in Chapter 3, equipment management has become a complex field dealing with complex problems, especially in high-tech industries in which maintenance management principles can no longer meet the needs. Therefore, this systems model is adapted to provide an effective framework for analyzing equipment management and the post-maintenance era. Figure 8.1 demonstrates

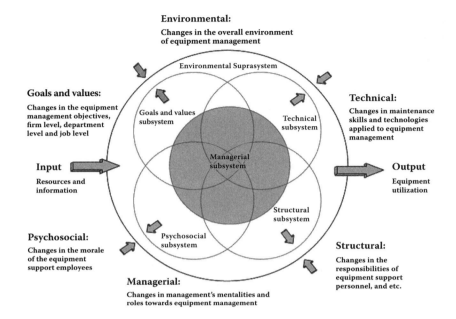

FIGURE 8.1
Systems model adapted for equipment management.

Kast and Rosenzweig's systems model with specific attributes applied to equipment management. In addition to examining the technical aspects of equipment management, the human factors and organizational structures are also analyzed in detail to develop a total systems view of the new equipment management approach. The intent is to present a broader view in equipment management and to tackle all pertinent equipment management issues from the systems viewpoint.

Equipment management is emerging as a complex field with issues in the technical arena, human factors, as well as organizational structures under prevailing environmental changes. This chapter uses the six key components from Kast and Rosenzweig's systems model as an outline to evaluate the history of equipment management. Figure 8.2 represents the framework used to analyze the facts and trends of each subsystem through the progression of the equipment management changes. Each of the subsystems within the model has key questions associated with it to guide the analysis. These key questions have an exploratory nature with the objective of promoting a thorough understanding of the issues and attributes of equipment management. Major measurable facets are defined and measured by sets of indicators to generate answers to these questions.

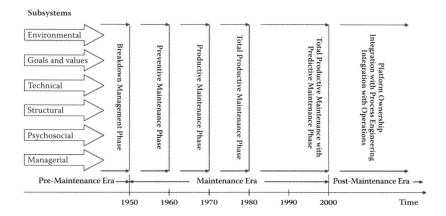

FIGURE 8.2
Framework for analyzing equipment management changes.

Based on analyzing the measurable facets, observable trends are apparent in all subsystems throughout the maintenance era. Also, considerable shifts in the trends are evident, with the exception of the environmental suprasystem, when moving into the post-maintenance era.

ENVIRONMENTAL SUPRASYSTEM

Equipment management is an important area of manufacturing organizations. At the macrolevel, it is operating under the same external environment as the entire organization. The fast-changing and dynamic characteristics of the business environment have the same impact on equipment management as they have on the other areas of the enterprises. New management practices and concepts, such as reengineering, drive changes from the top corporate level down to the successive levels of equipment operations management. In a similar fashion, new developments of business tools and methods, such as the Internet and communication technologies, create an enabling environment for the changes.

At the microlevel, because of their support function roles, equipment management personnel are under pressure to provide equipment services to the firm's internal customers. In addition to this customer-provider relationship, the operating environment of equipment management includes interrelations with other areas of the organization, such as planning, purchasing, facilities, and inventory control, as well as equipment manufacturers and vendors outside the firm. All interactions must be carefully examined when solving complex equipment problems.

The main theme of the environmental suprasystem in the systems model is examining the environmental changes that lead to the changes in equipment management objectives. These environmental changes have pushed equipment management into the post-maintenance era. There are two key questions in analyzing the environmental suprasystem: (1) What are the environmental changes that affect equipment management? and (2) How do these environmental changes stimulate the need to change equipment management approaches? To answer these questions, the interconnections between the key changes in the environmental factors and equipment management are illustrated in Figure 8.3.

Operational changes affect equipment usage and management at the highest level. Due to the recent technology developments, consumers expect high-tech companies to introduce new products into the market rapidly and to offer more choices as well. Therefore, product life cycle is shortened and product mix increases, demanding more engineering development efforts, so equipment must increase capabilities in supporting the operational changes. As a result, any new technology that can be utilized is incorporated into the equipment for the highest efficiency gains, which drives equipment changes.

Currently, equipment used in the high-tech industries has shorter life cycles, more capabilities, more complexity, and higher costs than previous

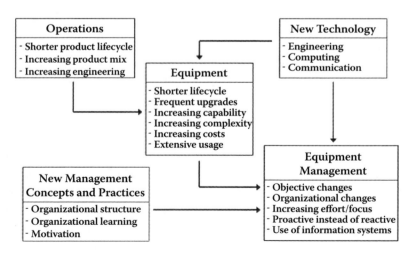

FIGURE 8.3
Environmental factors that have an impact on equipment management.

generations. This equipment is being used extensively, often around the clock 24 hours a day, 7 days a week. It requires frequent upgrades to keep up with the demands from new product introductions and engineering requirements. As equipment characteristics change, the traditional maintenance principles can no longer meet the needs described in Chapter 3. Therefore, equipment management approaches and practices must change.

Environmental changes associated with the availability of new technologies and new management principles also have significant impacts on the approaches applied to equipment management. For instance, the availability of database technology allowed the development of extensive equipment performance-tracking systems that enabled the predictive maintenance concept. Similarly, new management concepts, such as reengineering and total quality management, also affected the way equipment is managed.

In short, the equipment characteristic changes are initiated by the operational changes and fueled by the new technology developments. Similarly, the changes in equipment initiate the changes in equipment management, and the availability of new technology and new management practices continue to fuel the changes in equipment management approaches and practices. A summary of the environmental trends in equipment management is shown in Table 8.1.

GOALS AND VALUES SUBSYSTEM

The goals and values subsystem in the systems model provides a framework for examining the changes of goals and values under the environmental changes in the field of equipment management. The intention of understanding this subsystem is to set up the specific goals and values for measuring the success of the system. There two main questions associated with evaluating this subsystem: (1) What are the changes in equipment management goals and values from the corporate level to the departmental and individual levels? and (2) How do these changes in objectives and values prompt the change of equipment management approaches? In general, there are three major levels of objectives in a given company. The firm-level goal is at the highest level. The functional-level goals determine how different departments are working toward the corporate common goal. Last, the job-level objectives are based on how individuals work together to complete the tasks. Table 8.2 shows a summary of the objective trends in equipment management.

TABLE 8.1

Summary of Environmental Trends in Equipment Management

Key Area	Measurable Facet	Indicator	Trend
Operational scope	Production process changes	Number of major process changes	Increasing
		Frequency of major process changes	Increasing
	Product life cycles	Product introduction rate	Increasing
		Product obsolescence rate	Increasing
Equipment characteristics	Equipment complexity	Equipment capability	Increasing
		Equipment tolerance	Increasing
		Equipment module count	Increasing
	Equipment basics	Rate of equipment change	Increasing
		Equipment useful life	Decreasing
		Equipment acquisition cost	Increasing
		Equipment installation cost	Increasing
		Equipment maintenance cost	Increasing
New enabling technologies	New enabling technologies available to equipment management	Number of enabling technologies applied to equipment management	Increasing
New management concepts	New management concepts available to equipment management	Number of management concepts applied	Increasing

Reviewing the different equipment management phases described in Chapters 2 and 3, it is obvious that the equipment performance objectives change from one phase to another. In Tables 2.1 and 3.2, the most important goals of each equipment management phase were summarized. In fact, the goal differences are often the criteria that separate a particular phase from the others. For instance, only product defect and equipment breakdown rates were used in the preventive maintenance phase. In the productive maintenance phase, on the other hand, equipment management objectives included equipment reliability and maintainability. In addition to adding more indicators to represent the goals and values of equipment management from one phase to another, the magnitude of these indicators also changes from time to time. For example, a certain equipment breakdown rate would be acceptable in the preventive maintenance phase, but the objective of the total productive maintenance phase was zero breakdowns. The changes in goals and values can be utilized as indications to phase changes in equipment management, which certainly

TABLE 8.2

Summary of Objective Trends in Equipment Management

Key Area	Measurable Facet	Indicator	Trend (Maintenance Era)	Trend (Post-Maintenance Era)
Firm-level objectives	Change in corporate objectives	Corporate-level objectives with equipment-related content	Increasing	Increasing Alignment of objectives for all levels
		Joint departmental objectives on equipment performance	Increasing: More groups involved in equipment support	Decreasing due to centralized equipment support
Functional-level objectives	Change in charter	Department charter content	Increasing	Maintenance department has disappeared and process goals replace functional goals
	Change in equipment indicator reporting	Indicators generated and focused on by management	Increasing	Decreasing: Fewer maintenance indicators but more process and utilization indicators
Job-level objectives	Change in job position	Number of job positions	Increasing	Decreasing
	Change in job description	Job description content	Minimal change	Increasing and align with process objectives
	Change in roles and responsibility	Scope of responsibility	Minimal change	Increasing

signify the migration from the maintenance era to the post-maintenance era of equipment management as well.

Under the maintenance setup, the equipment performance objectives are mainly associated with the maintenance department. It discounts the different groups involved in equipment management that have their own goals and values that significantly affect the overall equipment performance objectives. The objectives vary depending on the organizational structure. In a typical organization, different departments are set up to support equipment directly or indirectly. As such, the departmental goals are different based on the functions performed by those departments. For example, most organizations have maintenance departments and separate equipment development engineering departments. In such an organizational setup, engineers making equipment selections tend to focus more on the capability and novelty of the equipment, and the maintenance department focuses on the reliability of the equipment. From time to time, this creates conflict because increasing capability and novelty generally means less reliability [16]. In simpler words, the more jobs a machine can do, the more parts are required, and the more things can fail. A machine that has 20 parts all operating at 99% reliability only has 82% reliability. The purchasing department, on the other hand, is under the general cost-saving principle, so the best equipment suitable for the jobs may not be purchased. The objectives of individuals within the departments are also different. Equipment repair technicians often focus on low-level routine repairs, and equipment engineers usually focus on nonfrequent and difficult equipment down incidents. The differences in their job functions often lead to differences in their objectives. Such issues have a great impact on the methods that a firm applies to manage its equipment. The overall goal cannot be achieved unless the goals and values of all involved parties are understood. Then, actions can be taken to unify these goals or to integrate these functions into a single process with a common goal, which is what the post-maintenance era intends to achieve.

In summary, the maintenance management discipline only focuses on the maintenance functional objectives, but equipment management includes other functions to establish a broader scope for effectively managing equipment. These functional areas include equipment development, purchasing, operations, planning, inventory control, logistics, and even parties outside the firm such as vendors and suppliers. Many of these areas have not been explored under the traditional maintenance approaches, but they are within the scope of equipment management under the systems model.

STRUCTURAL SUBSYSTEM

The structural subsystem is about how an organization is set up to manage equipment. It includes the defined responsibilities of each group associated with equipment acquisitions, maintenance, and operations. When a machine goes down, what is the first line of defense, and what are the response procedures? If the scope of a problem is beyond the floor repair personnel, what is the timeline for escalating the issue to higher levels and to whom? Who monitors the equipment, and who performs the routine maintenance? Who manages the spare parts, and who delivers them? Who generates indicators, and who analyzes them? What are the roles of vendors and equipment manufacturers, and how are their legal obligations bounded by the contracts? The structural subsystem in the organization determines the answers to these questions. Just like other systems, organizational structures do not stay the same.

As manufacturing equipment becomes increasingly complex, the structural subsystem becomes more complicated as well. For example, in the semiconductor industry, a chip-testing machine includes mechanical components, analog and digital circuits, and computer and network components. In addition to the hardware components, the machine can go down because of failures in many other elements, such as supply of compressed air, nitrogen, chilled water, electric power, room temperature and humidity, equipment control software, product testing programs, and even the product itself if it is defective on delivery. In many cases, a single repair involves repair specialists from more than one specialty field. From time to time, there are problems for which the root causes are not easily identified. All support groups have to work together to solve the problems, and such situations really put the equipment management structure to the test.

These increasing difficulty factors in equipment issues force the maintenance department and the other equipment support groups to work and interact differently. Under the maintenance setup, the approach to resolve a complex equipment issue is by putting more resources on it and getting more people involved. As a result, more people, departments, and cross-functional teams are drawn in to solve these issues, which increase the equipment support structure. The increase in organizational structure requires additional management coordination efforts and creates another set of issues, such as those related to teamwork and communication. In the post-maintenance era, the maintenance department disappears, and the maintenance function is either integrated upstream with equipment engineering or downstream with operations, so the ownership is clear and results in fewer parties being involved.

In examining the structural subsystem, three key questions guide the analysis: (1) Which organizational structural attributes, both internal and external to the company, have changed in equipment management?

TABLE 8.3

Summary of Structural Trends in Equipment Management

Key Area	Measurable Facet	Indicator	Trend (Maintenance Era)	Trend (Post-Maintenance Era)
Internal structure of equipment management	Formal organizational charts	Number of groups involved	Increasing	Decreasing
		Number of professions involved	Increasing	Minimal change
		Number of cross-functional teams	Increasing	Decreasing
		Number of employees	Increasing but leveling	Decreasing
	Departmental responsibilities	Number of functions performed	Minimal change	Focusing on process rather than function
		Resource allocations to functions	Increasing	No change to decreasing
External structure of equipment management	Vendor involvement	Number of vendor involved	Increasing slightly	Decreasing when integrating with engineering and increasing when integrating with operations
		Number of services performed by the vendors	Increasing	

(2) How do these structural changes have an impact on the changing of equipment management approaches; and (3) Which structural setup is considered successful and efficient in meeting the needs of the new business environment? The intention of analyzing the structural aspects of equipment management is to search for a better organizational structure in managing equipment. Table 8.3 shows a summary of the structural trends in equipment management.

TECHNICAL SUBSYSTEM

The technical aspect of equipment management involves two major areas. The first one is the technical skills of the people who deal with equipment. This aspect applies not only to the people who maintain the equipment, but also the experiences of equipment users have a major impact on equipment maintainability. Increasing the skills of the users can avoid many unnecessary down incidents. Traditional maintenance management approaches usually focus on the skills required for maintenance personnel. While this focus is an essential part of the technical requirements, the skills of management staff, planning staff, equipment selection engineers, and industry engineers are equally important. As equipment becomes increasingly complex, the required skills are more comprehensive and demand a greater amount of effort to obtain and maintain. Since many technologies are outdated rapidly, training becomes a continuous task. The equipment vendors and manufacturers are not exempted from such skill requirements, and their field service engineers must also go through skill retraining regularly to maintain the status of field experts.

The second technical aspect is the technologies used in managing equipment, such as computerized maintenance management systems (CMMSs) and predictive maintenance systems. Typically, computers are used to track equipment status and repair work orders. While some of the programs are off-the-shelf software application packages, some are custom developed for special purposes. Some are fully automated, and others require human interfaces. As the equipment management moves from one phase to another, more advanced technologies are used to manage equipment.

The three main questions that guide the analysis of the technical subsystem are as follows: (1) Which technical attributes, in both human skills and technologies applied to equipment management, have changed in equipment management? (2) How do these technical changes affect the equipment management approaches? and (3) Which technical basis is considered successful and efficient in meeting the needs of the new business environment? The first area of focus is to identify the change in skills required to manage equipment efficiently from one equipment management phase to another. For instance, the skill requirements of the maintenance personnel in the maintenance setup were equipment-specific skills such as electronic, electrical, and mechanical knowledge. The skill requirements for equipment support personnel in the post-maintenance era have extended to include general business skills such as presentation, negotiation, and project management skills.

Technology applied to equipment management is another area of focus. CMMSs were discussed extensively in Chapter 7. In the post-maintenance era, a new term, computerized equipment support systems (CEMSs) is used for the computerized solution used for equipment management. The difference between CMMSs and CEMSs is discussed further in Chapter 9. The key

TABLE 8.4

Summary of Technical Trends in Equipment Management

Key Area	Measurable Facet	Indicator	Trend (Maintenance Era)	Trend (Post-Maintenance Era)
Technical skill of equipment support and operations personnel	Individual job requirements and skills of maintenance personnel	Types of skill sets required	Increasing gradually	Increasing significantly
		Education level	Minimal change	Increasing
		Number of required training categories	Increasing gradually	Increasing significantly
		Training hours	Flat	Increasing
	Individual job requirements and skills of equipment operations personnel	Types of skill sets required	Increasing gradually	Increasing significantly
		Education level	Minimal change	Increasing
		Number of required training categories	Minimal change	Increasing
		Training hours	Flat	Increasing
Technology applied to equipment management	Computerized maintenance management systems (CMMSs) and equipment management systems (CEMSs)	Application of CMMSs and CEMSs	Increasing	Increasing with different design purposes
		Number of functionalities of CMMSs and CEMSs	Increasing	Increasing with focus on the entire equipment management process

distinction is their purpose: CMMSs are designed for maintenance managers, and CEMSs are designed for all personnel-related equipment, which includes the user community, as tools to help them do their jobs effectively. Table 8.4 summarizes the technical trends in equipment management.

PSYCHOSOCIAL SUBSYSTEM

Equipment is becoming increasingly complex due to advanced technologies used in the manufacturing processes. The demands placed on equipment support personnel are also increasing. Many of the traditional equipment support methods and structures result in too much reactive firefighting, which creates a stressful environment for everyone involved. Stress is a key source of employee morale problems, which eventually affect job performance. Equipment managers have a duty to provide a positive work environment through implementing proper systems and supplying necessary tools to the employees. Analyzing the psychosocial subsystem means taking a close look at the working environment and employee morale associated with equipment management personnel.

Similar to the other subsystems, three key questions are used to guide the analysis of the psychosocial subsystem: (1) What motivational changes have occurred to employees working with equipment? (2) How do these motivational changes affect the changing of equipment management approaches? and (3) Which employee motivation practices are considered successful and efficient in meeting the needs of the new business environment? The work environment has a large impact on employee motivations. As mentioned, the increase of equipment complexity has led to the increase of equipment issues and the difficulty factors in solving the issues. Since the maintenance setup is not capable of handling these issues effectively, it results in a stressful environment as more equipment excursion management efforts and firefighting are required. In the post-maintenance era, equipment issues are handled more effectively, and stress is reduced for the equipment support personnel. It also offers freedom and flexibility for the employees in performing their jobs, which produces a more positive working environment. Employee morale is another area of the psychosocial subsystem. Typically, it can be measured by employee satisfaction surveys, the amount of recognition employees receive, the number of projects and actions initiated by employees rather than management, the career advancement opportunities, and the employee career development programs. Table 8.5 presents the summary of psychosocial trends in equipment management.

MANAGERIAL SUBSYSTEM

The managerial subsystem plays a central role in bringing the entire organization to working toward a common goal. It has three major functions: planning, control, and integration. In planning, the inputs and outputs of the organization must be carefully examined and planned to achieve

TABLE 8.5

Summary of Psychosocial Trends in Equipment Management

Key Area	Measurable Facet	Indicator	Trend (Maintenance Era)	Trend (Post-Maintenance Era)
Work environment	Work-related stress	Number of excursions	Increasing	Decreasing
		Number of outstanding work orders	Increasing	Decreasing
		Amount of overtime	Increasing	Flat
	Freedom and flexibility	Decisions delegated downward	Remains low	Increasing
		Frequency of meetings with management	Increasing	Reducing
		Length of meetings with management	Increasing	Reducing
Employee morale	Employee satisfaction	Employee survey scores	Minimal change	Increasing
		Amount of employee recognition	Remains low	Increasing
		Amount of projects/ actions initiated by employee	Remains low	Increasing
	Advancement	Career development programs	Manager-dominated programs	Increasing scope plus employee participation
		Opportunities for job enlargement	Remain low	Increasing

the best profit results. In the field of equipment management, such planning includes the number of machines required, the number of support people required, desired equipment performance goals, and many other input-output factors. Furthermore, the managerial subsystem integrates all other subsystems in the model. Equipment managers acting within the managerial subsystem must understand the goals and values of each group related to equipment and integrate these groups to work toward a common goal. They must also understand the technical aspect of equipment management to select the best techniques to achieve the best results. Roles and responsibilities of the equipment management personnel must be clearly defined. Equipment managers must also promote a positive working environment and improve employee morale. The managerial subsystem is the backbone of the entire systems approach to equipment management.

When analyzing the managerial subsystem, three main questions are asked: (1) Which managerial attributes, in mindset and in actual practices, have changed in equipment management? (2) How do these managerial changes influence the changing of equipment management approaches? and (3) Which managerial practices are considered successful and efficient in meeting the needs of the new business environment? In Table 8.6, key measurable facets are listed. First, management's attention to equipment is evaluated on where equipment fits in the management priority list as well as on the level of management that pays attention to equipment. As the complexity and costs of equipment increase, it is not surprising that equipment management has been continuously climbing up the management priority list. In the maintenance setup, management is forced to do so because of the amount of excursions and firefighting. In the post-maintenance era, equipment management is still high on the priority list, but it is totally different in nature, providing robust equipment with increasing capabilities, as compared to the previous priority: fixing the equipment problems to keep operations running.

Second, the planning process changes are understood through examining indicators such as the number of elements or steps in the planning process as well as applications and functions in the planning tools. Under the maintenance setup, extra planning steps must be taken to reduce the impacts of equipment downtime, excursions, and development activities on operations. In the post-maintenance era, equipment maintenance is integrated with either development or operations so the planning activities can be consolidated; hence, the steps and parties involved in the planning process are significantly reduced.

Third, the supervision and control process changes are evaluated by examining indicators such as the type and the number of decisions made by the employees, the type and the number of directions given to the employees, the number of employees in equipment support, the type and the number of meetings run by managers, and the number of procedures and guidelines. There are obvious shifts from all these indicators from maintenance era to post-maintenance era, indicating less management supervision and control to the employees.

Last, level of integration is measured by the communication frequency between managers and the level of management participation in

TABLE 8.6

Summary of Managerial Trends in Equipment Management

Key Area	Measurable Facet	Indicator	Trend (Maintenance Era)	Trend (Post-Maintenance Era)
Management commitment to equipment management	Management attention	Equipment on management priority	Increasing due to excursions	Higher but not excursion driven
		Level of management attention	Midlevel unless in excursion	Higher but not excursion driven
Managerial practices in equipment management	Planning process	Number of elements in planning process	Increasing	Reducing and being delegated
		Application of planning tools	Increasing	Reducing
	Supervision and control process	Decisions made by employees	Remains low	Increasing
		Directions given to employees	Remains high	Decreasing
		Number of equipment support employees	Increasing or flat	Decreasing
		Meetings run by managers	Increasing	Decreasing
		Number of procedures and guidelines	Increasing	Decreasing
	Level of integration	Communication frequency between managers	Increasing between functions	Decreasing between functions and increasing between sites
		Management participation in equipment support teams	Increasing in cross-functional teams	Reducing in cross-functional team and increasing in cross-site teams

equipment support and problem-solving teams. Under the maintenance setup, maintenance managers find themselves increasingly involved with cross-functional teams and with other functional managers to solve equipment problems since they need help from these functions. In the post-maintenance era, equipment support ownership is consolidated, so equipment support managers have more control over the entire equipment platform and reduce the need for requesting help from other functions. Therefore, their participation in cross-functional teams is reduced. Instead, they increase their communication and involvement with cross-site equipment teams and managers to share and learn the best practices.

9

New Changes in the Post-Maintenance Era

INTRODUCTION

In Chapter 8, the descriptions of the systems model applied in equipment management briefly touched on the essence of the new changes and trends in the post-maintenance era. In this chapter, the important aspects of the post-maintenance era are explored and analyzed thoroughly to signify that it is indeed an important movement in the history of equipment. As revealed in the history of equipment management, the changing of the phases is typically initiated by environmental changes. As the existing management approaches and practices no longer meet the needs of the new environment, problems and inefficiencies surface. To fix these issues, new objectives are formed. In other words, the environmental changes impose changes on equipment management goals and objectives. To achieve the new goals and objectives, changes in the other subsystems follow, and a new phase takes form. As such, understanding the objective changes is the foremost step in comprehending the post-maintenance era.

EQUIPMENT MANAGEMENT OBJECTIVES

Driven by the environmental changes, equipment management objectives have fundamentally shifted since the late 1990s and early 2000s in many high-tech companies. As mentioned in the systems model description, there are three levels of objectives: the firm, the functional, and the job levels. The change in the firm-level objectives is minimal as they typically focus on business output and product development. Although equipment management objectives have not been explicitly stated in the corporate goals, it is believed that these objectives are implicitly reflected in the corporate-level objectives through the output performance objectives. Meeting the equipment management objectives is an important means to achieve the firm-level objectives. Regarding whether equipment objectives should be clearly stated at the corporate level, people hold different opinions. Some believe that firm-level objectives should be concise, focusing only on the end results desired by the company. Others believe that corporate objectives should be comprehensive and include major components such as equipment management. Whether equipment management objectives are explicitly stated at the firm level has a minimal impact on how equipment is managed. It is more important that the objectives in equipment management closely align with the firm-level objectives to enable the achievement of the desired corporate results. The main objective changes in the post-maintenance era are reflected at two levels: the functional level and the job level.

Functional-Level Objectives

Under the maintenance setup, the functional-level objectives of equipment management are solely reflected through the objectives of the maintenance department. The maintenance department is one of the functions in the functional organization setup. The functional viewpoint leads to the optimization of the function rather than the entire process. Applying the reengineering concept, equipment management should no longer be viewed as a function. Instead, it is an integrated part of the manufacturing process. The objectives associated with equipment management should not be called functional objectives. Instead, they should be called process objectives.

In the functional setup, the maintenance department is considered the owner of equipment performance. Throughout the phases as summarized in Chapter 2, the objectives of the maintenance function have progressed to higher levels and ultimately reached the highest goal of zero breakdown in the total productive maintenance (TPM) phases. However, the objectives of all the phases in the maintenance era remained with a common focus: equipment downtime—scheduled and unscheduled. The charter of the maintenance department is to minimize equipment downtime through continuous improvement of people skills and maintenance methods.

To measure performance against this charter, two groups of specific goals are used. The first group is associated with equipment performance indicators such as equipment reliability, availability, and maintainability. Since these measurements are equipment specific, each equipment type generally has its own performance goals. These goals generally start low for new equipment and are raised as the equipment matures. Eventually, these goals would reach the utmost level of zero downtime. The second group of goals is based on indicators that measure the effectiveness and efficiency of the maintenance methods, such as maintenance costs, average response time, and waiting for part time. These measurements are generally taken at the overall departmental level and are applied to the entire equipment base. Both groups of specific goals are functionally focused. Achieving these goals does not necessarily have an impact on the overall performance of the company.

Both groups of indicators may not have much practical value to the users. Here is an example of the user's perspective:

> I don't care if the equipment is up all the time and how much resources and effort you have spent to achieve that. When I planned a critical job using a particular machine, it better be functional at the time when I use it. My definition of functional means everything works. The equipment indicators only indicate that the machine itself is functional. In the case of facility and network problems, as a user, all I know is that I cannot use the machine. Telling me that the machine has no problem does not help me at all. And, I don't want you to send me around to see the facilities people or the IT [information technology] people.

In addition, certain operations may not require that the equipment is always available, and in many cases, immediate responses are not needed for certain down situations. Blindly putting the efforts to achieve goals such as 100% availability and zero response time is not an effective approach in managing such equipment. These goals are disconnected from the firm-level objectives.

Some equipment management practices in the maintenance era, especially in TPM, promoted the upstream interactions between the maintenance personnel and equipment design and development groups, as well as the downstream interactions with operational personnel. The bottom line is still focused on improving within the maintenance function but through receiving support from other functions. Therefore, these practices are often initiated by the maintenance department as requests for favors from these engineering and operations groups. All these groups have their own functional objectives; therefore, helping the maintenance function is often a low priority.

The main objective of a functional manager is for the business to grow, and that is how managers are trained to survive in the corporate world. In maintenance, growing the department means more headcount and budget, which must be justified by the workload. As such, the growing maintenance business means more downtime for the company. Conversely, if equipment performance is excellent and there is no downtime, the maintenance department may not exist, and the manager loses the job. Therefore, it is not to the benefit of the maintenance manager to improve equipment performance. This is the primary reason why the maintenance setup is ineffective. The approach to resolve this conflict is to align the departmental objectives with the company's output objectives by integrating the maintenance function into the factory's value-added process. Two directions can be taken: (1) integrating maintenance upstream with equipment development and (2) integrating maintenance downstream with production. For the upstream integration, the main departmental objective will be providing superior equipment capability and performance for the company, which includes achieving maintenance-free operation. For the downstream integration, the key departmental objective will be producing the best possible output, and achieving maintenance-free operation becomes a prerequisite. In either case, the maintenance department is gone. It is recommended that the upstream integration be used. Downstream integration does not provide a good career path for the maintenance workers, and it can be easily perceived as a demotion. It is not good for employee morale. As such, much of the following text is devoted to the upstream approach when discussing the post-maintenance era.

Moving into the post-maintenance era, there are several obvious changes in equipment management objectives. These objectives are transformed from the functional level to the process level, aligned more closely with the corporate objectives. With the disappearance of the maintenance department, a new department is formed to have a broader charter: "To develop and maintain a robust platform environment for the user community through upstream hardware development, cell innovation, and continuous improvement."

Under the maintenance setup, equipment repair personnel are released when the problems are determined to be unrelated to equipment. As manufacturing equipment becomes increasingly complex, the root cause of a problem may not be clear at first glance. For instance, equipment lockup could be caused by many things, ranging from equipment hardware failures, utility insufficiencies, and network issues to software bugs. Under the functional setup, there is no clear ownership of these problems, and the groups often point fingers at each other, leaving the users frustrated. Under the new setup, the equipment management personnel have the charter to provide "a robust platform environment" to the users. Their focus is no longer on equipment alone. Instead, it is the entire platform environment around the equipment. Hence, some organizations named the new department "platform engineering." Just meeting the equipment-specific goals is no longer considered sufficient. Providing the users with the right services at the right time with minimal effort is what counts the most.

Since the equipment-specific goals used to measure the maintenance function are inadequate, other objectives that are centered on the users' perspectives are added to measure the success of the new department. These goals have close alignment with the objectives of the equipment users' departments; in many cases, the same indicators are used, such as equipment utilization and product defect rate caused by the equipment.

Another addition to the new department charter relates to equipment engineering development. As mentioned in the environmental section of Chapter 8, frequent new product introductions and technology changes make equipment upgrades become common events. Consolidating equipment development and maintenance into a single ownership streamlines the equipment management process. It is a major change in organizational structure, and the details are discussed further in the next section on organizational structure changes. Goals and objectives are the focus of this section, and this new charter leads to the addition of new goals associated with equipment engineering development. Providing necessary equipment capability to the users becomes one of the objectives of the new department.

It is also necessary to mention that some equipment-specific objectives have changed in recent years in the high-tech industries. Extending equipment life used to be an important objective for the maintenance function as it saved capital investment and increased the firm's profit margin when the equipment was fully depreciated. However, technology has been changing so fast that equipment life ends by obsolescence of products rather than by irreparable breakdown. Now, machines are replaced either because they cannot produce the newer products or they are difficult to maintain since the technologies of these machines are old and no longer supported by the vendors. Therefore, extending equipment life is no longer a significant goal of equipment management in the post-maintenance era.

Besides, objectives based on statistical averages such as mean time between failures (MTBF) and mean time to repair (MTTR), which are used as important indicators measuring equipment reliability, gradually receive less attention. As mentioned, equipment is changing so frequently that there is hardly a long enough period to allow these statistical averages to become meaningful. When presenting these indicators, the numbers usually do not make sense, and the presenters have to explain and justify. The typical reasons are that the equipment is new, a new machine has been added to the fleet, or a new upgrade was performed. With these dynamic changes in the factory, these indicator numbers hardly show any trend and have minimal value.

In short, in the post-maintenance era, equipment-specific goals such as equipment availability, reliability, and maintainability receive less attention. Some may even be phased out. Instead, the goals that are closely aligned with the objectives of the users and the corporation are the center of attention. Also, these objectives span the entire platform environment throughout the equipment management process, which includes equipment utilization, equipment capability development, and sustaining equipment in all aspects (e.g., facilities, network, IT support, etc.).

Job-Level Objectives

Job-level objectives are associated with positions held by individuals in the organization. The job-level objectives naturally follow the objectives of the department under which the individuals work. The functional objective changes certainly reflect on the individual positions. Under the maintenance setup, there are two main technical positions within the maintenance departments: maintenance engineers and maintenance technicians. The objectives of the maintenance engineers are to develop and maintain the maintenance system and methodology. They determine the preventive maintenance (PM) schedules, the PM tasks lists, and the instructions and procedures on PM and repair tasks. The objectives of the maintenance technicians are to perform the PM and repairs by following the related instructions and procedures.

The job-level objectives of the maintenance functional setup have three typical characteristics. First, all of the objectives are associated with equipment maintenance only. Second, there is a functional division within the departments as engineers and technicians are responsible for different tasks. Again, the functional division leads to the optimization of the tasks but not the process as a whole. Accomplishing the job-level objectives does not mean the departmental objectives are achieved. Third, there is no accountability as the job-level objectives are shared by all employees holding the same positions. For instance, response to machine failures is by a pool of repair technicians. Many technicians may work on the same incident of downed equipment, especially when the repair lasts longer than a shift. If the repair does not go well or takes much longer than it should, there is no accountability regarding who is responsible for the problems or the delays. On the departmental level, if certain types of equipment do not meet the performance goals, it is difficult to determine whether the maintenance engineers or technicians need to improve their job performance. In other words, under the maintenance setup, the departmental objectives as stated in the previous sections belong to all employees of the department as a group, resulting in lack of accountability at the individual employee level.

Under the new platform ownership setup, all three characteristics have changed. First, equipment management activities are consolidated. Because the department has a broader charter and takes care of the entire platform environment of the equipment, the objectives of the individuals in the department go beyond equipment maintenance. Second, tasks are no longer divided by functions as each platform owner owns a particular equipment platform with all related functions. If the new tasks were divided by functions as before, using engineering development as an example, a new engineering group would be formed to perform that function, either by hiring new engineers or by transferring engineers who previously conducted the same function in different departments. This approach, however, would not correct the issues introduced by the functional setup. It just takes the issues under a single department. The job-level objectives for the engineering development personnel would stay the same and be based on job positions. Again, the functional setup often leads to a disconnection between departmental goals and the job-level objectives. In other words, the departmental objectives may not be accomplished even though all the job-level objectives are met.

To better align the individual job-level objectives to the departmental objectives, the platform engineering department allocates responsibilities based on equipment type. Individuals within the department, whether they are engineers or technicians, are assigned to one or more types of equipment platforms. Of course, engineers are assigned to more complicated and critical platforms, and technicians are assigned to less-sophisticated ones. Also, engineers often own multiple equipment platforms. these individuals are responsible for everything, complex or simple, associated with the equipment platforms that they own, from engineering development to sustaining and even equipment transfer. The job objectives for these individuals are the same as the departmental objectives except that the scope is smaller and limited to the specific equipment platform. It means that if all the individuals achieve their job-level objectives, collectively, the department achieves its overall objectives. Last, accountability is imposed on all individuals in the department. If a machine does not meet the performance goals or the users are not satisfied with the services from a specific equipment platform, the platform owner is the one who needs to take action to improve his or her performance.

ORGANIZATIONAL STRUCTURE CHANGES

In the structural aspect, the most significant distinction between the maintenance era and the post-maintenance era is that the organizational structure for equipment management emphasizes the overall process rather than the individual functions. Maintenance departments do not exist anymore, and maintenance, as a function, has disappeared and been absorbed in the company's value-added process. Often, all responsibilities associated with equipment management, from equipment introduction, to development, to sustaining are given to a single department. The new organizational structure is considered successful and efficient when demonstrated by headcount and downtime reduction as well as scheduling, output, and employee morale improvements seen in many implementation cases in high-tech companies.

During the maintenance era, the functional organizational setup is applied to equipment management. The maintenance department is the primary owner of maintaining equipment. As the operational scope and equipment characteristics change, there are increasing interdependencies between the maintenance department and other functional groups. Figure 9.1 shows a typical equipment management process under the functional organizational setup.

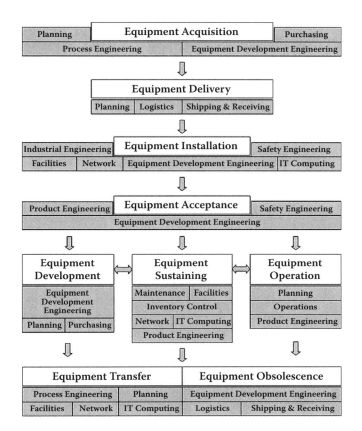

FIGURE 9.1
Equipment management process under the functional setup.

The process starts with equipment acquisition. Working together, process engineering and equipment development engineering groups decide on the types and configurations of equipment to be purchased. Matching the production or product engineering schedules, planning personnel decide on the timing and the quantity of the equipment to be purchased. Purchasing personnel then negotiate the pricing with the equipment vendors and generate purchase orders. The process then moves on to equipment delivery. Planning personnel are again involved to control the delivery schedule. Coordinating the transportation of the equipment from the vendor is a responsibility of the shipping and receiving personnel. Logistics personnel then take care of the decrating, rigging, and moving of the equipment onto the manufacturing floor.

The equipment installation step requires the industrial engineers to complete the layout, and then the equipment development engineers work with the vendors to complete the physical installation. Facilities personnel hook up all required utilities, such as electrical power, compressed air, water, and chemical lines. Network support personnel install telephones, network routers, lines, and other connection hardware. Finally, IT and engineering computing groups are responsible for installing the necessary software and configuring the computing environment. Safety engineers are also involved to make sure the physical layout of the equipment meets the safety standards and the installation process is conducted safely.

Equipment acceptance means performing equipment characterization and verification to determine if the equipment performance meets the required specifications. It often includes three major steps. First, the equipment must be tested to meet the equipment manufacturer's specifications. It often involves running the diagnostic programs that come with the equipment. The second step in acceptance is product verification. Real products or standards are used to verify the performance of the equipment. The third step is safety inspection, making sure that the equipment is operating under the federal and state Occupational Safety and Health Administration (OSHA) codes, and occupational hazards are documented and addressed with emergency response procedures. In a typical functional setup, equipment development engineers conduct the first step, product engineers complete the second, and safety engineers handle the last acceptance step.

Once the equipment acceptance is completed, the equipment is released to operations for product manufacturing or design and passed down to the maintenance department for sustaining. During the period when equipment is in use, many activities occur, and they can be grouped into three major areas: equipment development, equipment sustaining, and equipment operations. The operational phase also involves many functional departments. Equipment development engineers perform the upgrades and improvement projects. They also use the equipment for developing equipment capability for the future generation of products. Companies used to buy equipment just for engineering development, but as the costs of equipment increases, development must share equipment with operations to save costs. Maintenance personnel handle equipment repairs and PM tasks. Facilities and network support personnel respond to facility and network failures. IT and engineering computing groups cope with software issues and upgrades. Planning controls the timing and scheduling of all these activities to ensure that operations can meet product output goals and product engineers conduct product developments. In addition, purchasing is involved with the purchase of upgrade parts, spares, as well as services provided by the vendors. Inventory control manages the spare parts, tools, and materials. When equipment becomes obsolete or requires transferring or relocating, all the parties are involved again for its deinstallation and shipping.

There are many problems associated with the functional setup. First, there are frequent communication problems between the engineering departments, the maintenance department, and the operations departments. In many cases, equipment changes are not communicated clearly to the operations and maintenance personnel, making the maintenance and production tasks difficult. For instance, the equipment repair personnel as well as equipment users are often frustrated by the so-called Friday engineering releases (FERs), which means engineering upgrades or changes are typically released on Fridays, the last date of a regular workweek, because engineers try to meet project deadlines, which are often set by workweeks. The FERs are rushed out to meet the deadlines, leaving no time for thorough checkout or proper training. As a result, the operations and maintenance personnel struggle over the weekend without much engineering support. Communication can be a big problem when many groups are involved.

Another problem of the functional setup is that is causes conflicts between groups. The equipment management process involves many groups with different departmental objectives. For example, engineering departments focus on equipment capability, which may mean lower reliability and higher difficulty factors in maintainability. There are also numerous professions involved in the process, from technical personnel such as electrical, mechanical, and software engineers, to nontechnical staff such as planners, inventory control, and logistics personnel. These professionals come from different backgrounds and disciplines, bringing their own distinctive working styles and problem-solving approaches. Getting all of them to work together requires a great amount of effort. To deal with the communication problems and the conflicts between groups, many cross-functional teams are required, which is one of the distinctive characteristics of the functional organizational setup. It is typical that every project needs at least one cross-functional team. Employees complain that they cannot do much work as team meetings consume a good portion of the day. For instance, bringing new equipment into the factory is considered a big project in the functional setup. The project meetings need to include people from at least 10 departments shown in Figure 9.1.

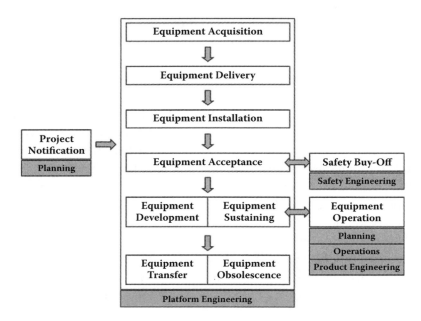

FIGURE 9.2
Equipment management process under the platform setup.

Clearly, these problems indicate that the functional approaches are not efficient in managing equipment performance. Each step of the process is performed by many functional groups. As a result, the process is broken into pieces with no accountability and clear ownership for each of the steps and the entire equipment platform performance. Since many parties are involved, communication problems often exist. The maintenance department, as the primary owner of equipment performance management, only plays a small role in the entire process. It cannot even take care of equipment sustaining alone. The ability of the department to provide adequate equipment performance is at the mercy of other groups.

To resolve the issues associated with the functional setup under the maintenance era, changing the organizational structure to the new platform ownership setup is the best solution. The entire equipment management process is streamlined by the platform engineering department. As shown in Figure 9.2, platform engineering owns the equipment management process from acquisition to equipment transfer and obsolescence. The planning department focuses on production schedules and capability planning models. Planning personnel only need to notify the platform owners when the users need the equipment and how much equipment is needed. The plans and schedules for the equipment purchase, delivery, installation, and acceptance are developed and managed internally within platform engineering based on the equipment release dates. The platform owners, acting as the project managers, control the schedule and the resources. The tasks, arranged and controlled by platform engineering personnel with schedules and costs in mind, may be done by other internal departments or outside contractors. Starting from purchase order generation to the final product acceptance, the entire process belongs to a single ownership.

The platform owners are also well trained in safety procedures. They are responsible for the safety issues associated with the equipment and the installation processes. They actually prepare the safety documentation for the safety buy-off, during which safety engineers spend only about 15 to 30 minutes checking the documentation and the equipment setup to make sure safety guidelines and OSHA codes are met.

Equipment developments and sustaining also belong to the platform engineering department. Under the maintenance setup, the time for knowledge transfer from equipment development engineering to maintenance is long, often months or even years. As new machines are frequently introduced and upgrades become common tasks, it takes a while for the maintenance personnel to complete the learning curve and obtain sufficient experience. With the same owner following the equipment since its acquisition, the knowledge transfer phase between development engineering and maintenance is eliminated. Equipment maintenance becomes an easier task as many issues may be encountered and addressed during the installation and acceptance process. With equipment becoming obsolete at a faster rate than ever before, this approach seems to be the best solution.

In addition, this setup also provides incentives for the platform owners to decline the acceptance of poorly performing equipment during the acceptance phase as they are the ones who need to take care of the equipment in operation. This approach addresses the common infancy failure patterns in modern equipment (see Chapter 1, Figure 1.2). Without the platform owner's signature on the equipment acceptance documentation, the equipment vendor does not receive full payment for the equipment. This new organizational structure gives platform engineering the power to hold the vendor accountable for any insufficient performance and services.

The new platform engineering structure also provides a clear and focused ownership. Bringing a new machine into the factory is no longer a big project. Equipment users are no longer passed around to see different groups, and there is no more finger-pointing between groups. Overall, this new structure promotes better services to the equipment users, namely manufacturing operators and product engineers from design and development groups. From the users' perspective, they have a one-stop shop for their equipment-related issues and training needs. The platform engineering department is their single contact.

As a result, the number of groups involved in equipment management is reduced significantly, from as many as 15 groups (as shown in Figure 9.1) in the functional structure under the maintenance era to the current 6 groups (as shown in Figure 9.2). This leads to a decrease in the number of cross-functional teams. Hence, less time is spent in meetings, which means more time is devoted to actual work. More important, communication issues and conflicts between groups are minimized. The equipment management process becomes simpler and easier to manage. Headcount is also significantly reduced as the maintenance activities are reduced, and the shift structure/supervisors are no longer needed. Task scheduling has also improved. Platform owners have control over their task schedules, making project timelines easier to meet. Many equipment development activities and upgrades can be done with scheduled PM, reducing the impact on the equipment users.

In summary, the new internal organizational structure, using the platform ownership concept, has successfully coped with the issues under the old organizational structure in the maintenance era. It provides an efficient equipment support infrastructure for operations and product development groups.

There are also changes in the external structure of equipment management that involve equipment manufacturers and support vendors. The platform owners now manage the relations with equipment manufacturers as well as the support vendors that provide equipment services such as fixture modifications, machining, custom-built parts, installations, rigging, and crating.

Under the functional organizational structure in the maintenance era, approved vendors are often established by the purchasing departments and sometimes with the involvement of engineering departments. Since there is no clear ownership of equipment management, doing a good job is seldom recognized, but making a mistake leads to much finger-pointing. Hence, people generally do not take risks and often choose medium to large companies to be on the safe side. Changing vendors requires too much effort and involves too many groups. When the platform owners gain clear ownership and receive credit for accomplishing exceptional results, they will be willing to take some risks to use new and often small vendors for cost saving. Also, the established vendors sometimes get overconfident by assuming business will continue to come their way, while small companies want the business so badly that they are willing to offer great services at a discount and with a quicker turnaround time. Therefore, when the platform owners gain the flexibility under the new organizational structure, they often capitalize on the money and time-saving opportunities by choosing new and small vendors. Sometimes, they even use these small vendors for jobs that used to be done by internal resources from other departments.

From the vendors' perspective, they also prefer the single-ownership idea as it also makes their jobs easier. The platform engineering setup provides a single point of contact for the vendors. Under the functional setup, service requests come from different groups without clear priorities. Some requests are even contradictory. Vendors also need to communicate changes to many groups. Now, the requests are presented to the vendors from one source with clear priorities, and the messages are clearly communicated both ways.

Another side benefit is that the quality of services provided by the vendors seems to have improved somewhat. The new platform engineering setup may take partial credit because it has the structure to hold the vendors more accountable for their equipment performance and services. Also, since the platform owners have the flexibility to change to vendors who provide better pricing and timing, this may create competition among the vendors and keep them on their toes for offering good-quality products and services.

THE PLATFORM OWNERSHIP CONCEPT

As the maintenance department transforms into the new platform engineering setup, the department has a new charter to provide total equipment support solutions. The individuals in the department become equipment platform owners. Their responsibilities cover the entire equipment life span, from initial equipment introduction to equipment obsolescence. Platform engineers and technicians coordinate equipment purchases with vendors and purchasing departments, perform equipment installations and acceptances, develop equipment support structures, negotiate service contracts and repair escalation support with the vendors, write equipment repair and PM procedures and specifications, and train equipment users to ensure the proper use of the equipment.

The platform engineers and technicians are empowered to decide how to manage the equipment that they own as long as the goals are met. Some platform engineers and technicians are technically oriented and decide to become content experts to solve most of the problems on their own. They put themselves on call 24 hours a day and 7 days a week and treat the equipment as their babies. Others operate in a different manner as they view themselves as administrators and area managers. They negotiate extensive support programs from their vendors. For instance, vendors are requested to provide 24/7 service support on new equipment warranties, and when performance goals are not met, the warranties will be extended with no additional cost.

Many platform engineers and technicians also focus on developing extensive repair procedures and training the equipment users so that the routine PM tasks and simple repairs can be done by the users. The job performance of the platform engineers and technicians is evaluated on the basis of the performance of the equipment that they own, not on how many hours and how hard they work. In the traditional maintenance departments, the job performance of the technicians is evaluated based on their technical skills and the number of repairs and amount of PM that they perform. In platform engineering, a technician may spend long hours and extensive effort to repair a machine, yet this approach is considered a negative performance issue since it is reactive. In theory, the platform owners' time should not be spent on repairs. Instead, their time should be spent on proactively ensuring equipment does not fail and on equipment development projects. They also have the responsibility to provide the necessary indicators to measure equipment performance. These indicators include equipment utilization in addition to availability, reliability, and maintainability. Working with others to implement the necessary computerized equipment management systems (CEMSs) to get these indicators is also one of their responsibilities.

The newly gained power comes with additional responsibilities. Figure 9.3 illustrates the roles and responsibilities of the platform engineering department. As mentioned, the departmental business scope is divided by the types of equipment platforms assigned to individuals as platform owners. Therefore, each individual in the department is

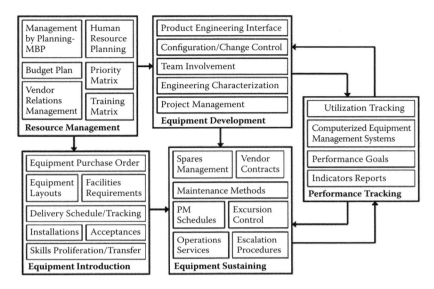

FIGURE 9.3
Platform engineering roles and responsibilities.

responsible for all five areas shown in Figure 9.3. To deal with business uncertainties, these platform owners are also assigned as backup owners to each other on one or more types of equipment platforms. Each individual is expected to manage equipment introduction, development, and sustaining on his or her primary owned equipment and, at the same time, to plan the resources and track the performance of the equipment.

The new organizational setup also enables an efficient cross-site equipment team structure for large corporations. The cross-site team structure includes equipment users groups (EUGs), equipment engineering teams (EETs), and equipment improvement teams (EITs) for a particular equipment platform. Platform owners who are in charge of that equipment platform at different factory sites attend these team meetings regularly. The EUG involves the owners of the platform at all sites and the equipment vendors. It is intended to be the single voice of the company, with collective power to bring issues to the vendors and request necessary services. The EET is an internal team in which all site platform owners share their best-known methods (BKMs) and general engineering methods (GEMs), prioritize improvement projects, and allocate developmental workload among sites. The EIT is a project team that focuses on a particular equipment improvement project. It may not involve owners of all sites and may include some technical experts from other departments or from vendors. Its function is to complete the equipment improvement projects and report the status to the EET. This cross-site team structure allows the platform owners to have forums to connect with others to obtain support and knowledge on the equipment that they own.

Compared to the previous maintenance approaches, the equipment platform owners have more responsibilities, more control over the equipment, and more flexibility in doing their jobs. As a result, the equipment management process has a clear and focused ownership. Employee morale has also significantly increased under the platform ownership concept because the employees can do what they prefer to do in pursuing their goals with greater flexibility, and the companies benefit from reducing headcount and equipment issues at the same time.

EMPLOYEE SKILL REQUIREMENTS

Under the functional organizational structure in the maintenance era, individual job titles are based either on the field of specialization (i.e., electrical engineers and mechanical technicians) or job functions (i.e., equipment development engineers and maintenance technicians). Despite the job title differences, job requirements are generally derived from disciplines or specialties. In other words, an engineer with an electrical engineering degree has a different job requirement compared to an engineer with a software engineering degree. Therefore, the job requirements are often limited within the disciplines in which the individuals are trained to perform. As a result, the technical skills possessed by individuals are often specialized and narrow.

Under the maintenance setup, electronic technicians in the maintenance departments only work on the electronic functions of equipment, while mechanical technicians take care of the mechanical functions of equipment. As equipment becomes more complex and includes parts from many disciplines, a specialized technician can no longer repair the equipment alone. Again, the traditional maintenance approaches are not successful and efficient in coping with current equipment issues. For instance, a mechanical specialist used to be able to fix machines independently that are mechanical in nature, but it is no longer the case as many mechanical movements in modern equipment are controlled by electronics and software. He or she then needs to wait for the electronic specialist or the computer specialist to solve the problems together. Obviously, this waiting time prolongs equipment downtime. Since several technicians are involved with the repair, the operations personnel may need to seek several people to get the status of the equipment and to understand how the machine is fixed. From time to time, the information is inconsistent as the specialists are responding from their field perspectives. The worst case is when the root cause is not clear. The mechanical technicians may conclude that the problem is electronic related, and electronic technicians may think that it is a software issue. Clearly, for maintenance personnel and operators, this setup makes their jobs harder, and it is a very inefficient way to manage complex manufacturing equipment.

To deal with the issues raised, some companies began to push for the universal tech concept in equipment maintenance departments. Maintenance technicians were required to acquire skills in multiple disciplines in order to respond to electrical, electronic, mechanical, and software issues, so that any equipment maintenance tasks and repairs could be handled by any one individual in the department. The job requirements were no longer discipline based. Training programs were developed to help technicians obtain a broader scope of skills.

While the maintenance technicians were going through the skills upgrade program, similar programs were being implemented in operations. In addition to operations skills, operators were required to obtain knowledge of the process and equipment. On receiving certification in these training categories, they gained the new title "manufacturing technicians." While running the equipment, these manufacturing technicians monitored production processes using statistical process control (SPC) tools. They were also expected to perform equipment cleaning, minor PM, and small repairs. The skill upgrade programs raised the technical skill requirements and standards in the maintenance and operations departments, leaving the qualified and proficient employees on the job.

The platform ownership concept, as a key step in transforming equipment management to the post-maintenance era, goes a step further than the universal tech concept. First, it affects engineers and put them under similar pressure in upgrading their skills. Second, a broader range of skills is added to the platform engineering job requirements. Additional skill requirements, such as project management skills, training skills, and leadership skills, are included in the job requirements of engineers and technicians. They also need to learn customer service and management skills. Under a platform organizational structure, as demonstrated in the

TABLE 9.1

Key Knowledge Requirements of the Platform Owners

Areas of Responsibilities	Required Knowledge
Resource management	Planning
	Budgeting
	Vendor relations
	Basic business laws
	Written and oral presentation
	Negotiation/business contracts
	Computer/office software
Equipment introduction, development, and sustaining	Safety policies and codes
	Basic building/layout knowledge
	Problem-solving techniques
	Electrical skills
	Electronic skills
	Mechanical skills
	Vision control technology
	Computer/server/network
	Operating systems: Windows/UNIX/Linux
	Product engineering
	Operations
	Project management
	Team building
	Leadership
	Customer services
	Training/teaching
Performance tracking	Statistical analysis
	Reliability theory
	Failure analysis
	Software design and programming

previous section, the platform engineering department ends up with a very broad scope of responsibilities. To perform the five areas of responsibilities shown in Figure 9.3, platform owners must possess the knowledge outlined in Table 9.1.

In the maintenance department, maintenance technicians typically have associate degrees and vocational degrees. When changing to the platform engineering department, only a few top maintenance technicians with excellent job performance have the potential to attain all the necessary skills. New hires with advanced degrees are often needed. Candidates with master's of business administration (MBA) and engineering degrees are preferred since the platform owners have responsibilities in both business and technical arenas. As a side note, the education level of the operations personnel also increases. Under the traditional organizational structure, operators are mostly high school graduates, with a few who finish vocational schools. While the maintenance organization is transforming, the manufacturing technicians in operations see this movement as a general trend for all employees to stay employed. They also start learning, and many go back to college while working. As a result, the overall educational level of the employees involved in equipment management has increased.

The broader scope of skill requirements is also reflected in the increased amount of training materials and training time. Under the maintenance setup, employee training is generally done in two categories. One is focused on the general orientation required for all employees of the corporation. Examples of these types of training are company policies, cultures, and safety procedures. Another category is focused on the specific functional skills required for an individual to perform his or her job. For instance, operators are trained in operations, and maintenance technicians are trained in equipment. Under the platform ownership setup, a formalized training matrix has been developed. In addition to the first category of general training, platform owners as well as manufacturing technicians are trained in five other categories: process, operations, equipment, training, and team, as illustrated in Table 9.2.

Each of these five categories has several proficiency levels. Each individual is expected to be certified at the minimum level of each training category. This gives each individual a broad perspective of the entire manufacturing process and serves as a foundation for further specialization of one or more categories based on the individual's job scope. Manufacturing technicians are certified on higher levels of operations and process training, while equipment platform owners receive in-depth training in equipment management. It is necessary to point out an obvious phenomenon: Training itself is a category in this training matrix. It indicates that training plays a key role in ensuring the success of the new approach in equipment management. Training is no longer a job for managers and the training department. It is everyone's job to share knowledge with coworkers. In fact, many organizations do not have training departments anymore, and the training function is integrated within the operational process.

With the diversity of skills, the platform owners are equipped to deal with various equipment issues in a timely manner. As the platform owners gain the knowledge of all components of the equipment, they can identify

TABLE 9.2

Training Categories for Employees Involved in Equipment

Training Category	Description	Example
Process	Knowledge and tasks associated with the manufacturing process and flow	Understand the manufacturing process and its critical elements Use of SPC tools to monitor the process Identify and correct nonstandard occurrences, follow procedures and specifications
Operations	Knowledge and sequential tasks associated with the process of running WIP (work in progress)	Machine operation safety Understand machine functions and operations Product handling Operate product analytical tools and record results Write and read passdown
Equipment	Knowledge and tasks associated with equipment management	Equipment safety Monitor equipment performance Perform PMs and repairs Installation and acceptance Development and modifications Equipment transfer
Training	Skills and knowledge associated with the proliferation of knowledge and information to fellow coworkers	One-on-one cross training Coaching Formal classroom instruction Training material development and revision Lesson planning
Team	Skills and knowledge associated with creating and maintaining a self-sufficient and effective work group	People skills Promote positive working environment Team building Leadership

the root cause of the problem and fix it in a short time without waiting for others to help. In addition, since the platform owners are trained in operations and processes, they understand the customers' needs, frustrations, and concerns, which allows them to serve the customers better. Sometimes when operations personnel are busy, the platform owners can actually help to run products.

From the operations side, manufacturing technicians are trained on equipment and are able to perform basic equipment PM and repairs. They now have a good understanding of the equipment and the frustrations of the platform owners in maintaining them. As a result, equipment failures induced by operator errors are reduced significantly. By understanding each other's business, the equipment platform owners and operations personnel work well together. Training also helps them to work together as an effective team. All of these things allow equipment performance to be managed efficiently.

COMPUTERIZED EQUIPMENT MANAGEMENT SYSTEMS

Under the maintenance setup, computerized maintenance management systems (CMMSs) are widely used to increase maintenance efficiency, but they are confined to the maintenance function. Moving away from the maintenance era, employee skill requirements are broader and multidisciplinary. CMMSs have been transformed into CEMSs that include functions outside the traditional scope of maintenance. CMMS is a standardized term used by maintenance professionals in all industries. The term *computerized equipment management system* is proposed in this book in conjunction with the introduction of the post-maintenance era. The major differences between the CMMSs and CEMSs are illustrated in Table 9.3.

As significant amounts of new computer and telecommunication technologies have been developed recently, functions and features of CMMSs (as discussed in Chapter 7) have significantly increased in all industries, but they are still focused on the maintenance function. As revealed in the history of equipment management, one of the key characteristics of the maintenance era is that equipment maintenance personnel are only responsible for availability. One of the major changes from CMMSs to CEMSs is that CEMSs have the capability of providing utilization indicators, which were previously out of the maintenance scope and not included in CMMS.

CEMSs represent a substantial change in computerized system design direction. CMMSs are mainly designed as tools for managers to forecast, plan, allocate, and control resources. Typically, the maintenance workers' interactions with CMMSs are limited to data entries. CEMSs are the tools developed for everyone involved in equipment, which includes the

TABLE 9.3

Comparisons between CMMS and CEMS

CMMS	CEMS
Computerized maintenance management systems	Computerized equipment management systems
Focus on maintenance functions—Improving equipment availability is utmost objective	Beyond maintenance—Total computerized solution for equipment management (includes utilization tracking)
Developed as tools for maintenance managers	Developed as tools for all personnel associated with equipment management
Focus on single-level tracking	Capable of providing multilevel tracking (cell/module, machine, and component-level tracking)
Designed for stable manufacturing environment (static cell configuration)	Designed with flexibility (dynamic cell configuration)

user community. That is why utilization and real-time equipment status capturing capabilities are important. Under the platform ownership setup, utilization data are no longer solely used by operations. The platform owners are providing total equipment support to the operations, and understanding the utilization of the equipment optimizes their efforts in maintaining the equipment and supporting operations.

Minimizing human interventions and efforts is another major change in CEMSs. Reducing human interventions means fewer mistakes in data entry and better data integrity. Since 2000, companies have been under economic pressure to reduce headcount. As a result, it is typical that an operator is generally responsible for several machines. Often, equipment operators are busy doing other tasks and may not notice a machine is down. In some cases, they may not have the ability to determine if a machine is down at all. For example, there have been incidents when equipment control screens did not indicate failure, but the machine was locked. Operators did not find this out until several hours later. Automated detection of equipment status is one of the new functions in the latest CEMS. In fact, some equipment manufacturers are including the automation feature as part of the machine software and offering it to customers without additional cost [17]. This feature is achieved by utilizing advanced software programming to monitor the computer central processing unit (CPU) activities of the equipment, the equipment error log files, and sometimes the network traffic to determine the status of the entire module. The utilization indicators are often provided through this method.

Automatic notification of equipment state changes is a function in CMMSs, as mentioned in Chapter 7. CEMSs take this feature to the next level and create greater benefits for the platform owners and operations. The automated dispatch systems are incorporated with the equipment-monitoring features to send messages to platform owners. In other words, the machine can actually inform the platform owners immediately when it has an issue, without any human intervention. With so many responsibilities placed on the platform owners' shoulders, this feature allows them to be able to multitask. This setup also significantly reduces response time and therefore reduces downtime. The use of this feature is not limited to down incidents. Platform owners can remotely access the equipment and monitor the equipment status. They can even perform maintenance tasks remotely without being physically next to the machines and often transparently without impacts on the users by running the checkout in the background. This feature also enables managers as well as product engineers to make decisions quickly without being near the equipment. It is extremely important for operations that involve multiple types of equipment operating 24 hours a day and 7 days a week.

Most CMMSs are developed under the assumption that a machine is a stand-alone entity. Therefore, they are only capable of single-level equipment tracking. Manufacturing technology has developed to a stage that most machines cannot perform a task independently. For instance, in the case of semiconductor testing shown in Table 3.1, the tester alone cannot perform the testing function. It requires other hardware, such as a test interface unit (TIU), a wafer- or chip-handling machine, and a control workstation. This combination of equipment setup is called a module or cell. CMMSs generally do not accommodate this level of complexity. Therefore, getting the equipment module and cell indicators is difficult. One of the important changes in CEMSs is to provide module-level, individual equipment-level, and even component-level indicators.

One might argue that a set of indicators on one level should be enough for managers to run the business. Most industrial engineering studies and literature possess this narrow view [9]. The multilevel equipment arrangement may have been unique in the past, but equipment designs have changed significantly in many industries, so a machine may not be able to produce output even though it does not have a problem. For operations managers, the output produced by the module is the focus, and the fact that the indicators have shown good performance on an individual machine is irrelevant. Module-level indicators are obviously needed.

On the other hand, machine-level indicators are also important. Equipment owners and manufacturers must know the performance of their machines to improve them. For instance, semiconductor testers and handlers are often made by different manufacturers because these machines are characteristically different in nature: One is electronically oriented, while the other is mechanical. The tester vendor has a contractual duty to meet certain performance goals, as does the handler vendor. Therefore, the ability to provide separate indicators for each machine through a CEMS allows platform owners to manage their respective vendors effectively.

One might also argue that cell-level indicators could be calculated by using equipment-level indicators; however, it is not that simple. In a simple case in which only two machines are connected to perform a task, downtime of equipment A plus downtime of equipment B may not be equal to the total downtime because both machines may go down at the same time. Or, neither machine may be down, but the interface between them goes down, whether it is a communication interconnection or a mechanical conveyor.

In some cases, a single machine is built with many components manufactured by different vendors. For instance, the semiconductor tester typically has a heat exchanger, power supplies, and circuit boards made by different vendors. Component-level indicators may also be required. One of the benefits of component-level tracking is allowing equipment platform owners to track broad upgrades and revisions easily. As mentioned many times, equipment upgrades become frequent tasks in manufacturing. New products typically require new revisions of parts. Often, a machine is installed with certain revisions of parts, but parts with newer revisions can be put into the machine during later repairs. Before the new product introduction, combinations of revisions may have been used since they do not affect the current generation of products. When a new product absolutely requires all new revision of parts, a great amount of effort could be spent to perform equipment audits to find out how many parts need to be ordered from the vendor. These audits often require that machines must be shut down and parts are pulled from the equipment. Hence, unnecessary downtime is introduced, and in some cases even failures are induced. With the component-level tracking feature, this information can be obtained in minutes without affecting normal operations.

A CEMS tracks equipment performance at several levels. These level-specific indicators enable the platform owners to break down root causes of failures, pinpoint the problems, and get the right manufacturers accountable for their issues. Therefore, CEMSs have become crucial enabling tools for achieving the platform ownership concept in the post-maintenance era. On the other hand, CMMSs used in the maintenance era failed to accommodate the increasing trend of multilevel modular setups in modern equipment.

Last, CMMSs are designed for stable environments. Most systems require an administrative right to add equipment or remove equipment. With a few CMMSs that have cell setup capability, the cell configurations are either hard coded in the software or require an administrator to change. In the recent business environment, equipment introduction and transfers are done much more frequently as product generation gaps are increasingly shorter. Within the factory, the cell configurations also change frequently as engineers are using the equipment for development as well. Using semiconductor testing as an example again, the TIU and some circuit boards are changed frequently for different product testing and engineering characterizations. The CMMS administrators are overwhelmed with these frequent equipment changes, and it takes time for equipment owners to submit the request and wait for the changes. The delay changes in the systems lead to data integrity issues, and the indicators are no longer accurate. CEMSs are designed with flexibility and allow frequent changes in equipment to be captured with minimal effort and even automatically. A dynamic cell-tracking methodology has been developed to provide multiple levels of indicators for the cells, the individual machines, and the components even if the individual machines and components are changed as frequently as possible [18]. The dynamic cell tracking is achieved by permitting user-defined configurations and using parent-child relations in database records to allow the separation of equipment components to be captured while maintaining the connections with the original configuration entry records.

WORK ENVIRONMENT CHANGES

As described, the functional organizational setup has no clear ownership of the entire equipment management process, which leads to finger-pointing and conflicts between groups. Within the functional departments, similar issues also exist between specialists. The finger-pointing and conflicts could bring about a hostile work environment if management does not pay attention and put forth timely efforts to contain them. The maintenance business deals with uncertainties, and people in general do not respond well to negative surprises. As such, frequent failures and excursions will certainly create a stressful environment for the employees and managers. As pointed out by behavioral management theories, stress is a key source of employee morale problems, which greatly affect job performance [19]. Overall, the equipment management approaches, using the maintenance function as the core, do not promote either teamwork or a positive work environment.

Furthermore, because many tasks require several groups and professionals from different disciplines to work together, it is often puzzling to associate failures or successes of these tasks to individuals. Measuring individual job performance is made difficult due to the lack of accountability. Hence, minimal actions can be taken to praise individuals for worthy achievements or to discipline the poor performers. As a result, employees seldom receive recognition. Working in a stressful environment without rewards for good performance, employees surely have morale issues.

The new equipment platform ownership structure reduces conflicts between groups and individuals. A positive work environment has been created as a result. In general, after the implementation of the platform ownership structure, managers and employees feel that there is lower tension, increased teamwork, higher morale, and increased job commitment in the new department. The platform ownership setup also provides positive conditions for employees to realize their self-actualization needs as stated in Maslow's human needs theory [19]. As mentioned, equipment platform owners have much more freedom and flexibility than maintenance technicians in the maintenance functional setup. They are free to choose work methods that fit their strengths as long as the equipment platform performance objectives are met. Some platform owners are working toward the technical career path and choose to become the content experts for all repairs and improvements. Others are working toward the managerial path to become area and program managers by getting the work done through vendors and other groups.

As a result, all decisions on the individual equipment platform level are delegated down to the platform owners. Managers no longer make decisions such as who works in relation to a down incident and which machine is repaired first. Managers are freed from micromanaging their employees' time and activities. Platform owners can decide their work schedule as long as the equipment is supported around the clock. Hence, these platform owners are operating under minimal supervision. Under the maintenance organizational setup, maintenance managers and supervisors meet with employees almost every workday to prioritize and allocate tasks. These meetings may take 1 to 2 hours depending on the workload. Under the platform ownership setup, platform owners generally meet with their managers once a week and only arrange additional meetings when necessary. The shift structure is no longer needed, and the shift supervisors are also eliminated. Management and administrative overhead can be reduced since the workers are becoming more independent.

The platform engineering setup also provides challenging work to the employees since the job responsibilities of the platform owners are extensive. To ensure that the platform owners have the abilities to succeed, comprehensive employee development programs are implemented to provide employees with the necessary job skills as well as self-enhancements. These programs include on-the-job training, off-site training, job rotations, and even on-site degree programs offered through partnering with universities and colleges. Tuition reimbursement is usually offered to all employees pursuing higher education in accredited universities, colleges, and vocational schools. Management's commitment to these programs is very high as employees are encouraged to participate. A mentality of continuous learning has been one of the by-products of implementing the platform ownership concept. The platform owners are also responsible for the training budget associated with their equipment platforms. They are empowered to arrange in-house or outside training classes for themselves and the users as needed. As a result, equipment management personnel find increasing self-fulfillment under the platform engineering setup in the post-maintenance era.

MANAGEMENT CHANGES

The managerial subsystem is the nucleus of all the subsystems. First, management must recognize the changes in the environmental suprasystem. Then, it needs to introduce and embrace all the changes in other subsystems. As described previously, the changes from the maintenance era to the post-maintenance era are substantial. If the managerial subsystem is not ready, none of these changes will be implemented successfully.

Under the maintenance setup, maintenance is considered an overhead function that does not directly produce profits. Therefore, maintenance is generally not an area of focus by management unless equipment performance is a gating factor for producing output. Whenever management is forced to cut budget or headcount, maintenance, among other overhead functions such as facilities and building services, would be on the top of the list for cost reduction. Consequently, equipment management is low on management's priority list, and the resources are consistently scrutinized for further reductions.

As capital equipment becomes increasingly expensive, it is noticeable that management's mindset toward equipment management has changed. Management starts to understand that equipment performance is crucial for product development and production. Many equipment management teams are formed at the corporate level with senior management involvement. Equipment management budgets have been increased with less scrutiny. It is easier for maintenance managers to justify resources for equipment performance improvement projects such as developing CEMSs. In short, equipment management has gained a great amount of attention from management, making the platform ownership concept a reality.

Managerial practice changes can be seen in three major functions of management: planning, control and integration. Under the maintenance setup, maintenance managers conduct planning, control, and integration at a very detailed level. The maintenance logistics described in Chapter 5 are the responsibilities of the maintenance managers. The first element in the planning process is labor planning. Maintenance managers keep track of the employees' work schedules and vacation time so that maintenance tasks can be thoroughly completed on each shift. The second element is equipment PM planning. Maintenance managers have to schedule the PM and communicate the schedule to the users so that the equipment can be freed at the planned maintenance time. Third, maintenance managers also need to conduct situational planning on responding to unscheduled down incidents. The fourth element is materials planning; maintenance managers plan the amount and types of spares and tools based on a historical run rate. The fifth element is budget planning, and the sixth element is planning for employee training.

Under the platform ownership setup, such detailed level planning, control, and integration have been passed down to the platform owners. Management's focus is on the high-level strategic planning that has a longer outlook and broader scope, such as departmental directions, change program implementations, employee development road maps, cross-process integration methodologies, and cross-factory integration that can further improve the equipment management process.

Supervision and control by management was reduced to a minimal level after the maintenance departments changed to platform ownership organizations. The platform owners make almost all the day-to-day decisions on equipment except in rare cases. The types of decisions made by equipment managers are often related to hiring, large-scale projects, and employee performance management. Furthermore, these managers do not need to micromanage the daily activities of the employees and give directions on what to do or how to perform specific tasks. Instead, managers provide help and directions on traditional management tasks, such as planning and vendor management, on an as-needed basis or when requested by the platform owners. As the employee independence increases, the number of subordinates reporting to a manager can be increased so the span of control can be larger.

Finally, the nature of integration by management has also changed as a result of moving away from the traditional maintenance organizational setup. Under the maintenance setup, maintenance managers frequently have to interface with other departments that have stakes in equipment; they communicate issues, request help, and resolve group conflicts. For instance, PM schedules need to be communicated to the operations and equipment development groups, and upgrade schedules need to be obtained from the equipment engineering development groups. With consolidation of the equipment management process, first there are fewer parties involved; second, task-oriented communication between departments is removed from the management level. Also, managers used to be involved in many cross-functional teams to resolve conflicts and issues. Now, these types of teams have almost disappeared, or if they exist, they are run and participated in by the platform owners instead.

This does not mean that integration is no longer a task of platform organization management. Platform engineering managers now take on a different type of integration. The integration under the maintenance organizational setup is between functional groups and generally within the same factory. Maintenance managers are often tied up with local tasks and operations, so that their management activities are restricted to their local sites. After the organizational change, equipment managers of large organizations regularly interface with other managers of other factories. In fact, ongoing teams of managers at different factories are established. These teams are often referred to as management committees (MCs). The cross-site equipment team structure mentioned with the EUG, EET, and EIT reports to the MC. The MC-EUG-EET-EIT structure is a critical part of the new changes since it provides three main functions. First, it is a communication forum for multiple factories to exchange information, from management practices to equipment repair methodologies. Second, it provides cross-site resource management by reducing duplicated efforts as well as sharing development project workloads. Third, it serves as a performance feedback channel for evaluating the job performance of the platform owners. This new level of integration by management enables and maintains a more efficient equipment management process for the entire company to create a virtual factory. Managers who work in smaller companies that have a single factory site will shift their efforts to interface with outside vendors and professionals in the industry, obtaining information and knowledge that can help the company be more efficient in managing the equipment.

SUMMARIZING THE POST-MAINTENANCE ERA

In summary, the post-maintenance era is the emerging phase of equipment management in which the approaches and practices in managing equipment are no longer centered on the maintenance functional setup. Instead, equipment management has a clear and focused ownership by consolidating functional tasks into a well-defined process. There are two general approaches in transforming away from the maintenance setup: upstream integration with equipment development or downstream integration with operations. The upstream integration, which practices the platform ownership concept, is highly recommended.

The transformation from the maintenance era to the post-maintenance era is prompted by the recent environmental changes in high-tech industries, especially by the significant increase in equipment complexity and costs as well as the shortening life cycle of equipment. These environmental changes lead to changes in equipment management objectives. The equipment management objectives in the post-maintenance era are no longer functionally based and are better aligned with the corporate-level objectives. The organizational structure for equipment management emphasizes the processes rather than the organizational functions. All responsibilities, from equipment introduction, development, and sustaining, are consolidated into a single department. Equipment personnel possess a broader range of technical as well as administrative skills. Advanced technologies are applied to equipment management in CEMSs. Equipment

TABLE 9.4

Characteristics of Platform Ownership Concept

Subsystems/Areas	Characteristics
Goals and values	Alignment of equipment management objectives with corporate and equipment users' objectives
	All individuals in equipment management have the same objectives applied to different platforms that they own
Structural	Consolidation of equipment management functions
	Single equipment management process ownership
	Equipment platform ownership
Technical	Broad range of skills possessed by employees
	Advanced CEMS and equipment support technologies
Psychosocial	High level of freedom and flexibility for employees
	Recognition programs that can be executed at the employee level
	Employee-oriented development programs
Managerial	Management does not micromanage tasks, focuses on providing strategic directions
	High-level and cross-factory integration
	Industrial relations, benchmarking, studies, and research

personnel also enjoy much more freedom and flexibility and as a result exhibit higher morale and self-fulfillment. Equipment managers typically focus on long-term strategic planning rather than tactical planning with higher levels of integration in equipment-related activities. Based on the systems model, the characteristics of the platform ownership concept implemented in the post-maintenance era are summarized in Table 9.4.

The maintenance era has lasted for a half century, and many organizations in many industries are still in it. The post-maintenance era captures the current state of equipment management in leading organizations in high-tech industries, and it is still in an early stage of development. Defining such an important phase provides a complete and up-to-date history of equipment management, as summarized in Table 9.5. The development of equipment management as a discipline will certainly continue to evolve and provide guidance to organizations to manage the increasingly complex and expensive equipment effectively as the technological revolution continues.

TABLE 9.5

History of Equipment Management Summary

	Period	Characteristics	Objectives	Concepts Developed
Phase 1				
Breakdown management	Pre-1950	Repair only when machines were down	Repair equipment failures in reasonable time	"If it ain't broke, don't fix it."
Phase 2				
Preventive maintenance	1950s	Establish maintenance functions Time-based maintenance	Extend equipment life Reduce unscheduled downtime and defects	Preventive maintenance Productive maintenance Maintainability improvement
Phase 3				
Productive maintenance	1960s	Reliability focus Maintainability focus Cost conscious	Reduce unscheduled downtime and defects while increasing maintenance efficiency	Reliability engineering Maintainability engineering Engineering economy Reliability-centered maintenance (RCM)

Phase		Date	Description	Target	Techniques
Phase 4					
Total productive maintenance (TPM)		1970s	Preventive maintenance plus total quality control and total employee involvement	Zero breakdowns and zero defects	Behavioral sciences Systems engineering Ecology Maintenance prevention Just in time (JIT) TQC and TQM Terotechnology
Phase 5					
TPM with predictive maintenance		1980s–1990s	TPM practices Condition-based maintenance Application of CMMS	Zero breakdowns and zero defects Optimization of availability	Computerized maintenance management Artificial intelligence and expert systems
Phase 6					
Post-maintenance era		2000s–	Total platform/process focus Functional integration: maintenance department vanishes Automation and advanced computing/communication applications	Optimization between availability and utilization Optimization of equipment development User satisfaction 100% maintenance free	Reengineering Circle of innovation Internet technology Instant messaging/text messaging Platform ownership Universal tech concept

10

Transformation and Implementation

INTRODUCTION

Changing from the maintenance era to the post-maintenance era requires a paradigm shift and substantial effort. The early cases of transformation from the maintenance era to the post-maintenance era took years to complete. These implementations were not smooth for three major reasons. First, the new equipment management approach was experimental, with no prior success examples. Second, it was a result of the increasing complexity of equipment issues and environmental changes. Therefore, during implementation, management was not able to focus solely on making changes and had to divide efforts to deal with the inherited problems from the functional maintenance approach. Finally, the changes were mostly initiated by the maintenance managers because of a long struggle with the perceived ownership but lack of total control of equipment. Obviously, the maintenance department had less power to make corporate changes as an overhead functional group. Managers have been through a long process of expanding the department charter by seizing upstream opportunities in equipment introduction and development. It had been a slow journey with consistent efforts in persuading upper management and management of other groups that previously owned equipment introduction and development. Based on reviewing the actions that have been taken in the pioneer cases, this chapter introduces a new transformation and implementation approach that will provide a better and quicker process to take an organization from the maintenance functional setup to the platform ownership setup in the post-maintenance era.

The transformation and implementation process, as revealed in Figure 10.1, may be used to transform equipment management from the maintenance era to the post-maintenance era. The first step is evaluating

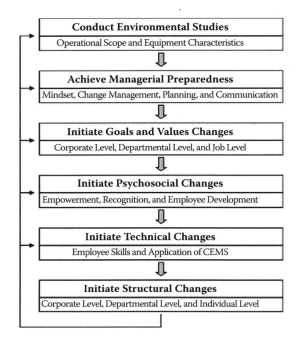

FIGURE 10.1
Implementation process for migrating to the post-maintenance era.

the environmental changes to see if there are needs to move away from the maintenance era. Future environmental trends should be taken into consideration so that proactive actions can be taken to avoid firefighting issues induced by environmental changes. The second step is preparing the changes at the management level, from mindset to resources. The third step is setting new objectives and values for employees and groups involved in equipment management. The fourth step is involving all employees to prepare them for the psychosocial changes. The fifth step is upgrading the skills of employees and implementing technologies as enabling tools to prepare the employees for their new roles. The final step is changing the organizational structure. This iterative implementation process should continue until the desired objectives have been achieved.

ENVIRONMENTAL STUDIES

Since different business environments have different requirements, it is extremely important to understand the business conditions before making any changes. The first area to be examined is the characteristics and scope of the operations. If the production process is stable with minimal changes and new product introductions are rare, then migration to the post-maintenance era may not be necessary. However, if there is a plan for major production process changes in the future, or there are desires to introduce several new product generations, the migration may help.

Whether to migrate essentially depends on equipment characteristics, the most critical environmental element to be studied. A short list of questions can be used to evaluate a given organization:

- Is new equipment frequently acquired?
- Does equipment introduction usually take more resources and time than expected?
- Are engineering developments and upgrades frequently required?
- Is equipment performance often below the desired goals?
- Do equipment users consistently complain about equipment issues?
- Do equipment issues generally require many groups working together to resolve them?

If the answer is "yes" to one or more of these questions, migration to the post-maintenance era should be seriously considered since it was developed to deal with these types of issues. In some cases, even if an organization has not yet experienced these types of issues, migration to the post-maintenance era may lead to headcount and equipment budget reductions. However, equipment complexity must also be considered since simple equipment would not justify the efforts to go through the changes, as headcount and budget savings may be insignificant.

MANAGERIAL PREPAREDNESS

The key to the success of any organizational change is management support. Once the decision is made to migrate to the post-maintenance era, management must start to prepare for the changes with the necessary priorities and resources. First, the traditional mindset, which viewed equipment management as an overhead function, must be changed. Migrating to the post-maintenance era not only saves costs and headcount but also speeds up product development and work-in-progress (WIP) turns and provides a safer and stress-free environment for the employees.

Second, management must select several key managers to act as change agents. These managers must be trained on organizational change management. Involvement and support from senior management is essential. Third, an implementation plan must be developed based on the current organizational structure and, more importantly, the culture of the organization. Companies with dynamic cultures may require less preparedness as employees are used to organizational changes.

Next, the key managers serving as change agents must communicate the purpose of the change as well as the implementation plan throughout the organization. Delivering organizational change messages is often not easy and deserves careful consideration. The message should be delivered to the departmental-level management first since their support is required to carry out the changes at the workers' level. A caution to the change agents is that resistance may be substantial at the department managers' level since one of the key elements of migrating to the post-maintenance era is for managers to relinquish tactical management tasks to the equipment platform owners. Not all managers are comfortable with giving up control. Frankly, one of the purposes for this organizational change is to reduce management efforts and headcount. Many functional groups may be consolidated during the change process. Furthermore, once the platform owners go through the change process and become self-sustaining, it is possible to eliminate the platform engineering manager and have the platform owners report to managers of other engineering groups, such as design engineering or product engineering. Upper management and the change agents must consider these issues and prepare plans to deal with them. Furthermore, managers may need to study organizational change management theories to gain a clear insight on how to handle organizational change issues effectively.

GOAL AND VALUE CHANGES

The first tangible change in migrating to the post-maintenance era should be on the goals and values of equipment management. A high-level objective for the entire equipment management process should be developed to serve as a charter for all employees involved in equipment management. The objective should be aligned with the corporate objectives and should focus not only on equipment availability but also on objectives such as customer satisfaction, equipment utilization, and equipment development.

These objectives should be quantifiable. An example of an objective statement could be to achieve 90% equipment utilization through the provision of vigorous equipment platform services and support.

During the implementation of changes in the organizational structure, detailed goals are needed to measure each department's progress toward achieving the desired changes. The scope of work is gradually increased, and the detailed goals are gradually added. Meeting all these detailed goals eventually means the migration is completed. For instance, while the maintenance department continues to operate under its functional objectives, a new goal associated with owning the equipment purchase process may be added as the first step. Next, an additional goal associated with certain engineering development tasks is included. Gradually, the maintenance department is acquiring more responsibilities and transforming to become the owner of the entire equipment management process.

Detailed goals at the job level are also needed to measure the change progress of the individual employees. For example, as the first step, employees may be asked to become certified in particular skills. Then, they are given the next objective of performing some planning activities on certain equipment platforms. Eventually, employees who successfully achieve all the given objectives become the platform owners. Employees who fail to meet the requirement would be transferred out of the department or put into the redeployment pool.

PSYCHOSOCIAL CHANGES

Many people do not like change, but if there are great benefits, they will be more willing and less reluctant to change. Before making any technical or structural changes, programs oriented toward employee motivation should be implemented. There are major benefits for doing so. First, these programs allow employees to realize that the change is good. They remove roadblocks and obstacles for the subsequent changes. Second, these programs become the basis for managers and even coworkers to reward good performers in the change process.

The first area of change toward employee motivation should be employee empowerment. Within the traditional maintenance departments, managers generally carry out all the planning and control activities. Under the platform ownership setup, platform owners are responsible for task management, including budgeting and project scheduling, which is a big change in employee empowerment. To achieve this objective, maintenance managers should free themselves from micromanaging detailed tasks, such as allocating routine work orders, scheduling preventive maintenance (PM), and so on. Instead, they should provide general directions and point out the desired results, empowering the employees to do whatever they see fit to achieve the results.

Employee recognition is another area that needs to be changed. A recognition system must be in place to reward not only good results but also, more importantly, good behaviors. Also, managers should not be the only ones to give recognition awards. Employees should be allowed to give recognition and rewards to other employees. This is extremely important for two main reasons. First, employees would truly realize the empowerment and feel that they are trusted. Second, it provides a significant support for the platform owners after the migration is completed. The platform owners are responsible for the entire equipment management process. As such, they often need to get the work done through others. For example, if a facilities worker works overtime to help a platform owner to meet an installation schedule or an operations worker takes the initiative to contain an equipment issue before it becomes a crisis, the platform owner should be able to reward them. The reward process should be simple and timely. The platform owner should not need to fill out too much paperwork or to seek many managers for approval. The rewards could be gift certificates, dinner coupons, or a small amount of cash.

The next area of focus is on employee development. As the platform ownership concept requires employees to obtain a broad range of skill sets, programs should be implemented to help employees gain new skills and knowledge. These programs should include offering tuition reimbursement for formal college degree programs and open-university courses, mentoring and coaching, job rotation and sharing, internal and external classroom training, and on-the-job training programs.

TECHNICAL CHANGES

Technical changes include two major areas: employee skills and technologies applied to equipment management. After putting the employee development programs in place, skill matrices and certification processes must be developed. The skill matrices should include the skill requirements for different job positions and grade levels. The certification processes may include a combination of tests, actual demonstrations of work, and job performance evaluations by managers and customers. Again, employees who successfully pass the certifications will become equipment platform owners, and employees who fail to master the skills will be transferred or redeployed.

While equipping the employees with new skills for the migration to the post-maintenance era, specific tools must be provided to make their jobs easier. Technology is the answer, and the main focus should be on computerized equipment management systems (CEMSs). Selecting a CEMS could be very confusing. Many commercial computerized maintenance management system (CMMS) software programs are available, but most of them are focused on traditional maintenance functions, such as work order tracking, labor management, and task scheduling. Also, most CMMS products are oriented toward classical manufacturing operations, such as steel mills and power plants, and are not suitable for the high-tech industries, for which the equipment base is under dynamic change.

Currently, there are no commercial CEMSs that meet all the needs of the post-maintenance era. In-house development or modification of existing systems is generally required. One of the approaches is to develop a CEMS from scratch, and the other is to select a commercial CMMS that meets the key requirements and develop it further to meet all the needs as pointed out in Chapter 9 under the CEMS section. Many CMMS vendors offer customization work to fit their existing programs to the users' requirements.

In addition to implementing CEMS, platform owners should also be equipped with mobile ability offered by telecommunication technologies. These tools include smart phones, notebook/netbook computers, and portable computing and planning devices such as personal digital assistants (PDAs) and tablets. All these tools greatly increase the efficiency of the platform owners and allow them to take on broader responsibilities.

STRUCTURAL CHANGES

The last step of the migration process is to change the organizational structure. A single group consisting of equipment platform owners should replace the maintenance department and other functional departments in the equipment management process. Each equipment platform should have primary and secondary owners assigned to provide for backups. For large, complex equipment platforms with a high equipment count, several owners may be assigned, but there should only be one primary owner as a team leader. For a simpler platform with a few pieces of equipment, the primary owner can also serve as a secondary owner of other platforms.

The organizational change should result in savings of human resources from the purchasing and planning departments since the platform owners take care of equipment development, purchasing, and equipment introduction. Therefore, these departments may have to go through headcount reduction processes. Some resources should be freed in other departments, such as process engineering, product engineering, industry engineering, and safety engineering, as the platform owners are now responsible for many activities that used to belong to these departments.

The maintenance and equipment development engineering departments are affected the most since they are consolidated under the new equipment management structure. The top performers from these two departments should be selected as candidates for platform owners. Employees from both groups need to go through extensive learning curves to become the platform owners. The maintenance technicians who do not succeed in the skill upgrade programs may be transferred to the operations department as manufacturing technicians. Equipment development engineers who do not meet the skill requirement may be transferred to other engineering departments, such as product engineering.

It is obvious that the organizational changes resulting from migrating to the post-maintenance era are significant. Many employees will change jobs, and some may lose theirs. It is important for management to offer programs such as early retirement and redeployment programs. Some employees may decide at the beginning to leave the company and not go through the extensive skill upgrade programs. Offering such programs at the beginning of the change process saves the company time and money in training employees who do not want to learn in the first place. Also, these employees may have a negative impact on the motivation of other employees.

The implementation process discussed is a general recommendation derived from case studies and my experience in implementing such changes in the high-tech industries. It may vary from one application to another due to different business conditions, structures, and organizational cultures. Additional actions or steps may be required to address specific situations. Overall, the migration from the maintenance era to the post-maintenance era is a paradigm shift in equipment management from many perspectives. The implementation process will not be easy, but the results are certainly worth the effort.

Glossary

availability: Time equipment can be utilized. It is often presented as a percentage of total time.

breakdown management: An approach in equipment management by responding to failures of an asset or a system as they occur. Phase 1 of equipment management.

cell or module availability: Availability of the entire equipment module. It is oriented toward operations/production and has a direct impact on output.

computerized equipment management systems (CEMS): Computerized systems applied to equipment management with features oriented toward managing equipment in a single process.

computerized maintenance management systems (CMMS): Computerized systems applied to equipment management with features oriented toward managing the maintenance function.

equipment availability: Availability of specific equipment only. It is oriented toward equipment owner and vendor management.

equipment cell or module: A group of individual equipment forming a unit or cell to perform the needed functions to produce output.

hard down: An infrequent equipment down incident that is difficult to repair.

maintainability: The average time period that the equipment was, or the probability that equipment will be, retained in, or restored to, a condition to perform its intended function.

maintenance: Combinations of all technical and administrative actions intended to retain an asset or a system in, or restore it to, a state in which it can perform the required functions.

maintenance management: A popular term that refers to the discipline of maintenance.

maintenance prevention (MP): An equipment management approach to eliminate equipment breakdown through designing or selecting equipment that is maintenance free.

mean time between failures (MTBF): A statistical average measuring reliability—how often the equipment fails. It is often used to predict equipment failure occurrence and forecast resources.

mean time to repair (MTTR): A statistical average measuring maintainability: how quickly the equipment can be repaired. It is often used to understand equipment failure occurrence as well as a measurement for repair skills; it also is used to forecast resources.

passdown: A shift-to-shift information handover or transfer for equipment status and progress on addressing issues, typical when there are issues that require longer than a shift to resolve.

platform ownership concept: An equipment management approach that converges all functional efforts into a single process with a single ownership. It is the new concept proposed in this book.

post-maintenance era: The equipment management era in which maintenance no longer exists as a stand-alone function.
Phase 6 of equipment management, proposed in this book.

predictive maintenance (PdM): A maintenance management approach that focuses on determining the life expectancy of components in order to replace them or service them at the optimum time.

preventive maintenance (PM): A maintenance management approach that performs a series of scheduled or planned tasks either to extend the life of an asset or a system or to detect critical wear that causes the asset or system to fail.
Phase 2 of equipment management.

productive maintenance: A maintenance management approach that applies reliability engineering, maintainability engineering, and engineering economy to optimize maintenance tasks.
Phase 3 of equipment management.

reliability: The average interval that the equipment could, or the probability that equipment will, perform its intended function within stated conditions.

reliability-centered maintenance (RCM): An analysis method that directs maintenance efforts at the parts and units where reliability is critical. It was started in the 1960s and was oriented toward aircraft maintenance.

Rock's law: The cost of capital equipment doubles about every 4 years in the semiconductor industry. A phenomenon observed and pointed out by a venture capitalist named Arthur Rock.

terotechnology: An equipment management approach developed in the United Kingdom and defined as a combination of management, financial, engineering, and other practices applied to asset management to achieve economic life-cycle costs.

total productive maintenance (TPM): A maintenance management approach that combines the American practice of preventive maintenance with Japanese total quality management and total employee involvement to achieve zero breakdown and zero defects. Phase 4 of equipment management.

universal tech concept: An equipment management approach by broadening the skill sets of the technicians so they can support all equipment or operations issues, as opposed to the specialist approach in which each technician is specialized in a particular field.

utilization: Time equipment is used. It is often presented as a percentage of total time. It was used as an indicator for operations/production only. It is an indicator for equipment management in the post-maintenance era.

References

1. Semiconductor Equipment and Materials International (SEMI). (2004). SEMI E10-0304 *Specification for Definition and Measurement of Equipment Reliability, Availability, and Maintainability (RAM)*. SEMI E10 Revision Task Force, San Jose, CA.

2. Campbell, John Dixon. (1995). *Uptime Strategies for Excellence in Maintenance Management*. Portland, OR: Productivity Press.

3. Moubray, John. (1999). *Reliability-Centered Maintenance*. 2nd ed. Oxford, UK: Butterworth-Heinemann.

4. Levitt, Joel. (1997, March). *Handbook of Maintenance Management*. New York: Industrial Press.

5. Nakajima, Seiichi. (1988). *Introduction to TPM: Total Productive Maintenance*. New York: Productivity Press.

6. Nakajima, Seiichi. (1989). *TPM Development Program: Implementing Total Productive Maintenance*. Tokyo: Japan Institute for Plant Maintenance.

7. Peters, Tom. (1997). *The Circle of Innovation*. New York: Knopf.

8. Cato, William W., and Mobley, R. Keith. (1999). *Computer-Managed Maintenance Systems*. Houston, TX: Gulf.

9. Peng, Kern. (2000, June). A new focus on equipment performance management. *Engineering Management Journal* 10(3), 156–160.

10. Peng, Kern. (2005, August/September). A new era dawns. *Manufacturing Engineer*, 84(4), 44–47.

11. Wireman, Terry. (1998, September). *Developing Performance Indicators for Managing Maintenance*. New York: Industrial Press.

12. Krajewski, L. J, Ritzman, L. P., and Malhotra, M. K. (2009). *Operations Management: Processes and Value Chains*. 9th ed. New York: Pearson.

13. Project Software and Development Incorporated (PSDI). (1999). *MAXIMO Information Kit*. Bedford, MA: Project Software and Development.

14. PSDI cited as leading worldwide supplier of maintenance management software by IDC. (1999, May 6). *PR Newswire*.

15. Kast, Fremont E., and Rosenzweig, James E. (1970). *Organization and Management: A Systems Approach*. New York: McGraw-Hill.

16. Gotoh, Fumio. (1991). *Equipment Planning for TPM: Maintenance Prevention Design*. Cambridge, MA: Productive Press.

17. Teradyne. *Automatic Time and Event (ATE) Logger—A Software Package for Logging and Reporting Tester Operations* [Company product information brochure]. San Jose, CA: Teradyne.

18. Peng, Kern. (2001, February). Advanced features of computerized equipment management systems. *SEMATECH Manufacturing Management Symposium*, SEMATECH, Austin, TX.

19. Shafritz, Jay M., and Ott, J. Steven (Eds.). (1996). *Classics of Organization Theory*. 4th ed. Orlando, FL: Harcourt Brace.

Endnote

Since the author has been teaching equipment management courses for many years, slide files for class teaching are available. Spreadsheet files for headcount and budget planning shown in Chapter 5 are also available. Please contact the author at kpeng@scu.edu to obtain these materials, as questions arise, or provide comments on the subject.

Index

A

ABC analysis, 135
acceptance, equipment, 12, 171–172
accessibility, 61–62, 136–137
accountability, 169
acquisition, equipment
    costs, 33
    functional setup, 171
    platform setup, 174–175
activities, 125, 130
adjustment, task example, 51
administrative labor and paperwork, 131
aging patterns, 13–14
AI, *see* Artificial intelligence (AI)
"airport" display feature, 139
alternatives, *xviii*
annual maintenance plan, 68–70
artificial intelligence (AI), 26
assessment form, 88, 89
assets, 7, 56–58, *see also* Equipment
assists, 11–12, *see also* Mean time between
    assists; Mean time to assist
assumptions, 68–69
audio alarms, 137
automated detection, equipment status,
    186
autonomous groups, 65
autonomous maintenance, 23–24
availability
    defined, 209
    equipment performance indicators,
        107–110
    equipment performance
        measurements, 14

B

background, equipment management, 3–5
backups, 206
bathtub curve, 13–14
beeping, low-volume, 137
best-known methods (BKMs), 91, 179

BKM, *see* Best-known methods (BKMs)
bottom-up budgeting, 81
breakdowns and breakdown management
    cost categories, performance
        indicators, 125–126
    defined, 209
    equipment management, 17–18
    goals and values, 150
    historical development, 27, 196
    maintenance, 11
    preventive maintenance program
        setup, 46–47
    productive maintenance, 24
    repair, 11
    TPM objectives, 23, 63
broadband analysis, 59
budget planning, 75–81, 192, *see also*
        Planning and budgeting
business environment
    enabling technologies, 33–35
    equipment characteristics, 31–33
    equipment management, 29–36
    fundamentals, 29–30
    management concepts, 35–36
    operational changes, 30–31

C

calendar-based trigger, 47, 49
calendar module, 135–136
calibration
    equipment/tool requirements, 54
    preventive maintenance, 4
    task example, 51
Campbell, John Dixon, 55, 63–64
categories, cost performance indicators,
        125–126
Cato, William W., 127
cell availability, 209
cell-level indicators, 188
CEMS, *see* Computerized equipment
        management systems (CEMS)
certification matrix, 92

challenges, annual maintenance plan, 69
change agents, 202
change orders, purchase agreements, 98
changes, annual maintenance plan, 68
changes, post-maintenance era
    computerized equipment management
        systems, 185–189
    employee skill requirements, 180–185
    equipment management objectives,
        163–170
    functional-level objectives, 164–168
    fundamentals, 5, 163
    job-level objectives, 168–170
    management changes, 191–194
    organizational structure changes,
        170–177
    platform ownership concept, 177–180
    summary, 194–197
    work environment changes, 189–191
checklists, *see* Task lists
chemical analysis, 58–59
clean, *see* TLC process
cleaning, task example, 51
clean rooms, complexity, 32
CMMS, *see* Computerized maintenance
        management systems (CMMS)
color codes, 83
color lights, 137
communication, *see also* Notification
    change agents, 202
    computerized maintenance
        management systems, 137–138
complaint ratio, 121
complexity
    equipment, 3
    semiconductor manufacturing, 31–32
    structural subsystem, 153
components, maintenance prevention,
        61–62
computer-based technology
    computer-based training, 91
    database software programs, 25–26
    disk operating system, 136
    equipment trends, *xviii*
    hardware, 37
    inventory management, 102–103
    preventive maintenance, 4
    software, 37, 80
    Web technology, 34, 91

computerized equipment management
        systems (CEMS)
    changes, post-maintenance era,
        185–189
    CMMS comparison, 185–189
    defined, 209
    platform ownership, 178
    technical subsystem, 155–156
computerized maintenance management
        systems (CMMS)
    accessibility, 136–137
    calendar module, 135–136
    CEMS comparison, 185–189
    communication, 137–138
    customization, 140–141
    data entry, 123, 138–139
    equipment module, 132
    errors, 123
    features, 136–141
    financial module, 135
    flexibility, 140–141
    functions, 132–136
    fundamentals, 5, 127–128
    implementation, 141–142
    integration, 139–140
    inventory management, 103
    inventory module, 134–135
    labor module, 134
    notification, 137–138
    objectives, 128–131
    presentation, 138–139
    preventive maintenance module, 133
    safety module, 134
    security, 136
    structural inefficiency, 38–39
    technical subsystem, 155–156
    TPM with predictive maintenance,
        25
    work order module, 132–133
Computer-Managed Maintenance Systems,
        127
consistency of assets, 7
contract management
    budget planning, 79
    maintenance management logistics,
        95–100
control process changes, 159
conversion, equipment setup, 11
corrective action, 10

costs
    behavior, equipment maintenance, 21
    breakdown by categories, 125–126
    equipment, 3–4
    equipment acquisition, 33
    holiday preventive maintenance
        schedule, 50
    life cycle, 66
    per activity, 125
    per equipment, 124–125
    performance indicators, 124–126
    preventive maintenance, 19
    productive maintenance, 20–21
    rates, 124–125
customer services and management
    communication, 130–131
    customer request/complaint ratio, 121
    maintenance management logistics,
        92–95
    satisfaction, 121–122
customization, 140–141
cycle counting, 103–104

D

database software programs, 25–26
data entry, 123, 138–139, 186
data review, task example, 51
defects
    productive maintenance, 20, 24
    TPM objectives, 23, 63
defragmentation, 4
delivery, equipment
    functional setup, 171
    platform setup, 174–175
*Developing Performance Indicators for
    Managing Maintenance,* 105
development, equipment
    functional setup, 171
    fundamentals, 12
    platform setup, 174–175
development of employees, *see* Skills and
        skill level; Training
DHTML, *see* Dynamic HTML (DHTML)
diagnosis and diagnostics
    preventive maintenance, 4
    repair time, 10
    task example, 51
    vs. fixer, 53

direct costs, 126
disk operating system (DOS), 136
dispatch systems, 187
documentation
    information system, 54–55
    task example, 51
DOS, *see* Disk operating system (DOS)
downtime
    expectation and acceptance, 21
    per shift, 119
    post-maintenance era, 164
    problems in objectives, 38
downtime/headcount ratio, 119
dynamic HTML (DHTML), 34

E

educational degrees, 182–183
EEP, *see* Excursion escalation procedures
        (EEPs)
efficiency, productive maintenance, 20
electrical monitoring, 60
e-mail feature, 138
employee skill requirements, *see* Skills and
        skill level
enabling technologies
    business environment, 33–35
    environmental trends, 150
encrypted data, 136
engineering economy, 20–21
engineering state, 8
engineering time, 10
environmental incidents, 107
environmental studies, 200–201
environmental suprasystem, 147–149
equipment
    actions, 11–12
    availability, 209
    cell or module, 209
    changing characteristics, 3
    characteristics, 31–33, 150
    complexity, 3
    condition trigger, 49
    costs, 3–4, 124–126
    downtime, 10
    environmental factors, impact, 148
    failure patterns, 12–14
    life span, 3–4
    module, 132

performance measurements, 14–15
planning, 192
preventive maintenance, 53–54
purchase agreements, 96, 97
safety indicators, 106–107
setup, 11
states, 8
test, repair time, 10
time, 8–11
tool requirements, 53–54
uptime, 9
equipment-dependent availability,
108–109
equipment engineering teams (EETs),
179
equipment management
background, 3–5
breakdown management, 17–18
business environment, 29–36
environmental factors, impact, 148
framework, change analysis, 146–147
fundamentals, 5, 7–8
historical developments, 17–27, 42,
196–197
issues of maintenance, 36–40
maintenance management
comparison, 6
maintenance phase summary, 26–27
post-maintenance era, 29–42, 197
predictive maintenance, 25–26
premaintenance phase summary,
26–27
preventive maintenance, 18–19
productive maintenance, 20–22
terminology, 8–15
total productive maintenance, 22–26
equipment management objectives
functional-level objectives, 164–168
fundamentals, 163–164
job-level objectives, 168–170
equipment management process, systems
view
environmental suprasystem, 147–149
fundamentals, 145–147
goals and values subsystem, 149–152
managerial subsystem, 157–161
psychosocial subsystem, 156–157
structural subsystem, 152–154
technical subsystem, 154–156

equipment performance indicators
availability indicators, 107–110
maintainability indicators, 113–115
reliability indicators, 110–113
safety indicators, 106–107
utilization indicators, 115–117
equipment users groups (EUGs), 179
errors, *see* Operational misses and error
rates
escalation structure, 84
excursion escalation procedures (EEPs),
84
expectations
changes, annual maintenance plan,
68
downtime, 21
equipment users', 94
expert systems, 91
extensible markup language (XML), 34
external structure, trends, 154

F

facility charges, 80
factory acceptance, 12
fail-safe backup, manual labor as, 17
failure
logs, 54
modes, reliability-centered
maintenance, 56–57
patterns, equipment, 12–14
repair, 11
unscheduled failure reduction, 18
fast-changing environment, *xvii–xviii*
fatality cases, 107
features, CMMSs, 136–141
feedback, 121
FER, *see* Friday engineering releases (FER)
figures, list of, *xiii*
file backup, 4
financial module, 135
firm-level objectives
post-maintenance era, 163–164
trends, 149, 151
first-aid cases, 107
fixed costs, 126
fleet indicators
availability, 110
maintainability, 114–115

reliability, 112
utilization, 117
flexibility, 140–141
format and variation
  availability indicators, equipment,
    108–109
  cost breakdown by categories, 126
  cost rates, 124–125
  customer satisfaction indicators, 121
  labor productivity indicators, 118–119
  maintainability indicators, equipment,
    113–114
  nonproductive downtime indicators,
    120
  operational misses and error rates,
    122–123
  reliability indicators, 110–111
  safety indicators, equipment, 106–107
  utilization indicators, equipment,
    116–117
formulas
  CMMS data entry error rate, 123
  Cost per (activity), 125
  Cost per equipment, 124
  Customer request/complaint, 121
  Downtime %, 109
  Downtime/headcount, 119
  Downtime/shift, 119
  Escalation rate, 122
  Headcount/equipment, 18
  Mean Cycles Between Failures, 112
  Mean Time Between Assists, 111, 112,
    115
  Mean Time Between Failures, 110, 112,
    113
  Mean Time Off-Line, 113, 114–115
  Mean Time To Assist, 113, 114
  Mean Time To Repair, 113–114, 115
  Mean Units Between Failures, 111,
    112
  PM misses rate, 123
  Prime Time Utilization, 116
  Spare outage rate, 123
  Supplier Dependent Availability, 109
  SWAT rate, 122
  Weekly Dependent Availability, 108
  Weekly Manufacturing Availability,
    108
  Weekly Operational Availability, 108

Weekly Operational Utilization, 116
Weekly Total Availability, 108
Weekly Total Utilization, 116
Work order/headcount, 18
Work order/shift, 119
forums, information sharing, 51
frequency
  maintenance prevention, 61–62
  preventive maintenance schedule,
    47–50
Friday engineering releases (FER), 173
functional-level objectives
  post-maintenance era, 163–168
  trends, 149
functions
  computerized maintenance
    management systems, 132–136
  identification and loss of, 56

G

garbage in, garbage out (GIGO), 138
general engineering methods (GEMs), 91,
  179
general maintenance
  equipment/tool requirements, 53–54
  fundamentals, 5, 45
  information system, 54–55
  maintenance prevention, 61–62
  materials/parts requirements, 54
  predictive maintenance, 58–61
  preventive maintenance, 45–55
  reliability-centered maintenance,
    55–58
  schedule, 47–50
  staffing requirements, 52–53
  task lists, 50–52
  terotechnology, 65–66
  total productive maintenance, 63–65
  triggers, 49
GIGO (garbage in, garbage out), 138
goals and values, *see also* Objectives
  annual maintenance plan, 68
  platform ownership characteristics,
    195
  systems view, equipment management
    process, 149–152
  transformation and implementation,
    202–203

H

Handbook of Maintenance Management, 58, 65
Hard down, 209
hardware upgrades, 37
hazardous materials, 134
headcount
    planning and budgeting, 70–75
    problems in objectives, 37–38
headcount/equipment ratio, 118
high-tech equipment needs, *xvii–xviii,* 5
historical developments, *xvii–xix,* 5, 17–27, 196–197
holiday preventive maintenance schedule, 50
HTML, *see* Hypertext markup language (HTML)
human resources, 134, 206
humidity, rooms, 32
hypertext markup language (HTML), 34

I

idle state, 8
idle time, 10
IM, *see* Instant messaging (IM)
implementation, 141–142, *see also* Transformation and implementation
indicators
    availability, 107–110
    cell-level, 188
    environmental trends, 150
    maintainability, 113–115
    managerial trends, 160
    objective trends, 151
    psychosocial trends, 158
    reliability, 110–113
    safety, 106–107
    structural trends, 154
    technical trends, 156
    utilization, 115–117
    value of, 168
indirect costs, 126
"infancy mortality," 13–14
information, sharing, 51

information system, 54–55, *see also* Recording-keeping practices
input unit based trigger, 48, 49
installation, equipment, *see also* Transformation and implementation
    functional setup, 171
    fundamentals, 12
    platform setup, 174–175
instant messaging (IM), 138
integrated circuit chips, 31–32
integration, *see also* Transformation and implementation
    computerized maintenance management systems, 139–140
    management changes, 192–194
    managerial subsystem, 159–160
intellectual property (IP), 87
internal structure, trends, 154
International Organization for Standardization (ISO) 9000, 54
Internet and intranet technology, 34, 41
interrupt, assist, 12
Introduction to TPM, 22, 63
inventory management
    maintenance management logistics, 100–104
    reducing storage, 131
inventory module, 134–135
IP, *see* Intellectual property (IP)
issues of maintenance
    fundamentals, 36–37
    problems in objectives, 37–38
    structural inefficiency, 38–39
    unsuitability, 39–40
"it-is-not-my-job" mentality, 24

J

job-level objectives
    equipment management objectives, 168–170
    post-maintenance era, 163–164
    trends, 149
just in time (JIT) methodology
    inventory module, 135
    total productive maintenance, 23, 65

K

Kast, Fremont E., 145–146
knowledge-based systems, 91
knowledge requirements, 182

L

labor module, 134
labor planning, 192
labor productivity and indicators,
    118–119
LAN, *see* Local-area networking (LAN)
Levitt, Joel, 55, 58, 65
licenses, software, 80
life cycle
    CMMS implementation, 141
    maintenance prevention, 61
    terotechnology, 66
life span, equipment, 3–4
local-area networking (LAN), 26
lockout procedures, 134
lodging expenses, 79
logistics
    budget plan, 75–81
    contract management, 95–100
    customer services and management,
        92–95
    fundamentals, 5, 67
    headcount plan, 70–75
    inventory management, 100–104
    planning, 67–85
    strategic planning, 68–70
    suppliers, 95–100
    tactical planning, 82–85
    training and people development,
        85–92
    vendors, 95–100
lost-day cases, 107
lubrication, 51, *see also* TLC process

M

machine condition based trigger, 48, 49
machine cycle trigger, 48, 49
maintainability
    defined, 209
    equipment performance indicators,
        113–115
    equipment performance
        measurements, 14
maintenance
    defined, 6, 18, 209
    era, 26–27
maintenance management
    comparison, *xvii*, 6
    defined, 6, 209
    equipment management comparison,
        xvii
    high-tech equipment needs, *xvii–xviii*
maintenance management logistics
    budget plan, 75–81
    contract management, 95–100
    customer services and management,
        92–95
    fundamentals, 5, 67
    headcount plan, 70–75
    inventory management, 100–104
    planning, 67–85
    strategic planning, 68–70
    suppliers, 95–100
    tactical planning, 82–85
    training and people development,
        85–92
    vendors, 95–100
maintenance performance indicators
    availability indicators, 107–110
    costs, 124–126
    customer satisfaction indicators,
        121–122
    equipment, 106–117
    fundamentals, 105–106, 117
    labor productivity indicators, 118–119
    maintainability indicators, 113–115
    nonproductive downtime indicators,
        119–120
    operational misses and error rates,
        122–123
    processes, 117–123
    rates, 124–125
    reliability indicators, 110–113
    safety indicators, 106–107
    utilization indicators, 115–117
maintenance phase, 26–27
maintenance prevention (MP)
    defined, 209
    general maintenance, 61–62
    historical developments, *xvii*, 6

maintenance system
   budgeting, 75–81
   computerized maintenance
      management systems, 127–142
   concepts and practices, 45–66
   contract management, 95–100
   cost performance indicators, 124–126
   customer services and management,
      92–95
   equipment performance indicators,
      106–117
   fundamentals, 5
   inventory management, 100–104
   maintenance management logistics,
      67–104
   maintenance prevention, 61–62
   people development, 85–92
   performance indicators, 105–126
   planning, 67–75, 82–85
   predictive maintenance, 58–61
   preventive maintenance, 45–55
   process performance indicators,
      117–123
   reliability-centered maintenance,
      55–58
   supplier management, 95–100
   terotechnology, 65–66
   total productive maintenance, 63–65
   training, 85–92
   vendor management, 95–100
management changes, 191–194
management concepts
   business environment, 35–36
   environmental factors, impact, 148
   environmental trends, 150
managerial preparedness, 201–202
managerial subsystem, 157–161, 195
manual labor, as fail-safe backup, 17
manufacturing availability, 108
manufacturing time, 9
Maslow's human needs theory, 190
Material Safety Data Sheets (MSDS),
      134
materials/parts requirements, 54, 83
MAXIMO, 139
MCBF, *see* Mean cycles between failures
      (MCBF)
mean cycles between failures (MCBF),
      14, 111

mean time between assists (MTBA)
   reliability indicators, 14, 111
   training and development, 87
mean time between failures (MTBF)
   availability indicators, 107
   defined, 209
   minimal value, 168
   productive maintenance, 20–21
   reliability indicators, 14, 110
   training and development, 87
   unsuitability, 39
mean time off line (MTOL), 15, 114
mean time to assist (MTTA)
   maintainability indicators, 15, 114
   training and development, 87
mean time to repair (MTTR)
   availability indicators, 107
   defined, 210
   maintainability indicators, 15,
      113–114
   minimal value, 168
   productive maintenance, 20–21
   training and development, 87
   unsuitability, 39
mean units between failures (MUBF), 111
"micron" naming convention, 30
microprocessor segment as driver, 30–31
Microsoft Windows environment, 34
migration, 200, *see also* Transformation
      and implementation
mission statement, 68
Mobley, R. Keith, 127
model, systems, 145–146, *see also* Systems
      view, equipment management
      process
module availability, 209
morale, 157–159
Moubray, John, 55, 58
MP, *see* maintenance prevention (MP)
MSDS, *see* Material Safety Data Sheets
      (MSDS)
MTBA, *see* Mean time between assists
      (MTBA)
MTOL, *see* Mean time off line (MTOL)
MTTA, *see* Mean time to assist (MTTA)
MTTR, *see* Mean time to repair (MTTR)
MUBF, *see* Mean units between failures
      (MUBF)
musical tones, 137

N

Nakajima, Seiichi, 22–23, 63
"nanometer" naming convention, 30
nonproductive downtime and indicators
    fundamentals, 10
    indicators, 119–120
    process performance indicators,
        119–120
non-recurring engineering (NRE) costs,
    87
nonscheduled state, 8
nonscheduled time, 8
notification, *see also* Communication
    automated, 187
    computerized maintenance
        management systems, 137–138
    fast, 129
    platform setup, 174–175

O

objectives, *see also* Goals and values
    annual maintenance plan, 68
    computerized maintenance
        management systems, 128–131
    issues of maintenance, 37–38
    maintenance prevention, 61
obsolescence
    functional setup, 171
    platform setup, 174–175
    post-maintenance era, 167
    rates, operational changes, 30
Occupational Safety and Health
        Administration (OSHA), 172,
        173
off-hour preventive maintenance, 50, 83
office computing technology, 26
oil analysis, 59
on-call support infrastructure, 80
on-the-job training, 191
operational availability, 108
operational changes, 30–31
operational misses and error rates,
        122–123, 186
operational scope, 150
operational utilization, 116
operations
    environmental factors, impact, 148

    equipment, functional setup, 171
    platform setup, 174–175
    research models, 20
    time, 8
operator mentality, 24
opportunities, annual maintenance plan,
    69
organizational setup
    maintenance department, 36, 40
    post-maintenance era, 40
    problems in objectives, 38
    structural inefficiency, 38–39
    structure changes, 170–177
OSHA, *see* Occupational Safety and
        Health Administration (OSHA)
output quality based trigger, 48, 49
output unit based trigger, 47–48, 49
overhead function
    annual maintenance plan, 69
    structural inefficiency, 38–39
    traditional mindset, 201
overhead time, 9
ownership
    platform engineering setup, 175–176
    post-maintenance era, 41

P

paperwork reduction, 131
Parkes, Dennis, 24
particle analysis, 58–59
parts
    budget planning, 79
    kits, 54
    maintenance prevention, 61–62
    nonproductive downtime indicators,
        120
    replacement logs, 54
    waiting, nonproductive downtime
        indicators, 120
passdown, 210
passwords, 136–137
patterns, failure, 12–14
payroll costs, 81
PdM, *see* Predictive maintenance (PdM)
Peng, Kern, *xxiii*
people development, 85–92
performance indicators
    availability, 107–110

maintainability, 113–115
reliability, 110–113
safety, 106–107
utilization, 115–117
period-specific utilization, 116–117
personnel, *see also* Headcount
annual maintenance plan, 70
maintenance pools, 53
nonproductive downtime indicators, 120
universal tech *vs.* specialist, 85–86
Peters, Tom, 36
phases
equipment failure pattern, 12–13
fundamentals, 4
historical developments, 196–197
pre-1950s (breakdown management), 17–18, 27, 196
1950s (preventive maintenance), 18–19, 27, 196
1960s (productive maintenance), 20–22, 27, 196
1970s (total productive maintenance), 22–25, 27, 197
1980s-1990s (total productive maintenance with predictive maintenance), 22–25, 27, 197
2000s (post-maintenance era), 40–42, 197
planning
accurate cost data, 131
budget plan, 75–81
fundamentals, 67–68
headcount plan, 70–75
management changes, 192
process changes, 159
strategic planning, 68–70
tactical planning, 82–85
platform, 5
platform engineering setup
job-level objectives, 169–170
post-maintenance era, 40–41, 167
roles and responsibilities, 179
platform ownership
changes, post-maintenance era, 177–180
characteristics, 195
defined, 210

job-level objectives, 169
work environment, 190
PM, *see* Preventive maintenance (PM)
pools, maintenance personnel, 53
post-maintenance era
business environment, 29–36
computerized equipment management systems, 185–189
defined, 210
employee skill requirements, 180–185
environmental studies, 200–201
environmental suprasystem, 141–149
equipment management, 29–42, 196–197
equipment management objectives, 163–170
fundamentals, 5, 29
goals and values, 149–152, 202–203
introduction, 40–42
managerial changes and subsystem, 157–161, 191–194
managerial preparedness, 201–202
new changes, 163–197
organizational structure changes, 170–177
platform ownership concept, 177–180
psychosocial subsystem and changes, 156–157, 203–204
structural changes and subsystem, 152–154, 205–207
summarization, 194–197
systems view, equipment management process, 145–161
technical changes and subsystem, 154–156, 204–205
transformation and implementation, 199–207
work environment changes, 189–191
predictive maintenance (PdM)
capability, 128
defined, 210
equipment management, 25–26
general maintenance, 58–61
historical developments, *xvii*
technical subsystem, 155
predictive maintenance (PdM) with TPM, 25–27
premaintenance phase and era, 26–27
preparation, 109–110

presentation
    computerized maintenance
        management systems, 138–139
    cost breakdown by categories, 126
    cost rates, 125
    customer satisfaction indicators, 122
    labor productivity indicators, 119
    maintainability indicators, equipment,
        114–115
    nonproductive downtime indicators,
        120
    operational misses and error rates, 123
    reliability indicators, equipment,
        110–111
    safety indicators, equipment, 107
    utilization indicators, equipment, 117
preventive maintenance module, 133
preventive maintenance (PM)
    cost breakdown by categories, 126
    defined, 210
    equipment management, 18–19
    equipment/tool requirements, 53–54
    fundamentals, 12, 45–47
    goals and values, 150
    historical developments, *xvii,* 27, 196
    information system, 54–55
    materials/parts requirements, 54
    operational misses and error rates, 123
    planning, 192
    schedule, 47–50
    staffing requirements, 52–53
    task lists, 50–52
    triggers, 49
prioritization, 82
procedures, *see* Task lists
processes
    customer satisfaction indicators,
        121–122
    fundamentals, 117
    labor productivity indicators, 118–119
    nonproductive downtime indicators,
        119–120
    operational misses and error rates,
        122–123
process performance indicators
    customer satisfaction indicators,
        121–122
    fundamentals, 117
    labor productivity indicators, 118–119

nonproductive downtime indicators,
    119–120
operational misses and error rates,
    122–123
process planning, 83–84
production setup, 11
productive maintenance
    defined, 210
    equipment management, 20–22
    goals and values, 150
    historical development, 27, 196
productive mean time between failures,
    110–111
productive state, 8
productive time, 9
product run time, 9
product verification, 172
profit, 3–4
psychosocial changes and subsystem
    platform ownership characteristics,
        195
    systems view, equipment management
        process, 156–157
    transformation and implementation,
        203–204
purchase agreements, 95–100
purpose
    availability indicators, equipment, 107
    cost breakdown by categories, 125
    cost rates, 124
    customer satisfaction indicators, 121
    labor productivity indicators, 118
    maintainability indicators, equipment,
        113
    nonproductive downtime indicators,
        119
    operational misses and error rates, 122
    reliability indicators, equipment, 110
    safety indicators, equipment, 106
    utilization indicators, equipment,
        115–116

R

rates, 124–125
RCM, *see* Reliability-centered
    maintenance (RCM)
reactive problem-solving
    breakdown management, 18

productive maintenance, 21
psychosocial subsystem, 157
staffing requirements, 53
real-time feedback, 121
recording-keeping practices, 130, *see also*
    Information system
reengineering
    fundamentals, 7
    management concepts, 35
reliability
    defined, 210
    equipment performance indicators,
        110–113
    equipment performance
        measurements, 14
reliability-centered maintenance (RCM)
    defined, 210
    general maintenance, 55–58
    historical developments, *xvii,* 6
    productive maintenance, 22
    steps, 22
reorganizing, task example, 51
repair
    breakdown maintenance, 11
    breakdowns, maintainability
        indicators, 114
    cost breakdown by categories, 126
    fundamentals, 11
    productive maintenance, 20
    time, 10
replacement, task example, 51
reports
    management concepts, 35
    providing indicator, 129–130
    real-time, 139
resource scheduling, 83
response flow charts (RFCs), 84
response time, 10, 120
responsibilities
    job-level objectives, 169–170
    platform engineering, 179
    platform ownership, 182
RFC, *see* Response flow charts (RFCs)
risks and risk factors, 69–70, 176–177
Rock's law
    defined, 210
    equipment acquisition cost, 33
    semiconductor industry, 3, 29
Rosenzweig, James E., 145–146

S

safety
    equipment performance indicators,
        106–107
    equipment performance
        measurements, 14
    module, CMMS, 134
    platform setup, 174–175
    purchase agreements, 99
schedules and scheduling
    down state, 8
    downtime, 10
    preventive maintenance, 47–50
    resources, 83
security, 136, *see also* Safety
self-learning, 91
semiconductor industry
    business environment changes, 29–30
    enabling technologies, 35
    equipment characteristics, 31–33
    machinery need, 17
    operational changes, 30–31
    post-maintenance era, 40
    Rock's law, 3, 29
    structural subsystem, 153
setup time, 11
short life cycle, 33, 87–88
shutdown
    nonscheduled time, 8
    procedures, 134
signaling, communication/notification,
    137
site acceptance, 12
situational planning, 192
skills and skill level, *see also* Training
    changes, post-maintenance era,
        180–185
    educational degrees, 182–183
    matrices, 204–205
    platform ownership, 182
    staffing requirements, 53
    technical subsystem, 155, 156
"smart" machines, 25
software
    licenses, budget planning, 80
    upgrades, problems in objectives, 37
SOP, *see* Standard operating procedures
    (SOPs)

spares
  budget planning, 79
  effective management, 129
  inventory management, 100
  operational misses and error rates, 123
  purchase agreements, 98
specialist personnel, 85–86
spreadsheets
  budget/equipment plan, 76–78
  headcount plan, 71
stable environments
  annual maintenance plan, 69
  of assets, 7
  computerized maintenance
      management systems, 189
  traditional practices, 39–40
staffing requirements, 52–53
standardization, 61–62
standard operating procedures (SOPs), 84
standard run, 10
standby state, 8
standby time, 9–10
start time, 47–50
startup time, 8–9
status, automated detection, 186
stockroom, inventory management,
      101–102
strategic planning, 68–70
stress, 157, 201
structural changes and subsystem
  platform ownership characteristics,
      195
  systems view, equipment management
      process, 152–154
  transformation and implementation,
      205–207
structural inefficiency, 38–39
supervision process changes, 159
supplier-dependent availability, 109
suppliers, 95–100, *see also* Vendors
support
  budget planning, 80
  nonproductive downtime indicators,
      120
survey ratings, 121
sustaining, equipment
  functional setup, 171
  platform setup, 174–175
Suzuki, Tokutaro, 23

SWAT escalation
  operational misses and error rates, 122
  tactical planning, 84
systems view, equipment management
      process
  environmental suprasystem, 147–149
  fundamentals, 145–147
  goals and values subsystem, 149–152
  managerial subsystem, 157–161
  model, 145–146
  psychosocial subsystem, 156–157
  structural subsystem, 152–154
  technical subsystem, 154–156

T

tables, list of, *xv*
tactical planning, 82–85
task lists, 50–52, 82
technical changes and subsystem
  platform ownership characteristics,
      195
  transformation and implementation,
      204–205
technology
  environmental factors, impact, 148
  telecommunication technologies, 205
  vendors, lack of support for old, 168
telecommunication technologies, 34
temperature
  monitoring, 59
  rooms, complexity, 32
terminology, equipment management,
      8–15
terotechnology
  defined, 24–25, 210
  fundamentals, 65–66
testing integrated circuit chips, 31–32
test interface unit (TIU), 187
text messaging, 34, 138
The Circle of Innovation, 36
tighten, *see* TLC process
tightening, 51
TIU, *see* Test interface unit (TIU)
TLC process (tighten, lube, and clean), 45
tools planning, 83
total availability, 108
total productive maintenance (TPM)
  defined, 211

equipment management, 22–24
functional-level objectives, 164
general maintenance, 63–65
goals and values, 150
historical developments, *xvii,* 6, 7, 27,
    196–197
with predictive maintenance, 25–26
problems in objectives, 37
total productive maintenance (TPM) with
    predictive maintenance
equipment management, 25–26
goals and values, 150
historical developments, 27, 197
total quality control (TQM), 22–23,
    65
total utilization, 116
TPM, *see* Total productive maintenance
    (TPM)
TPM Development Program, 22, 63
TQM, *see* Total quality control (TQM)
training
    budget planning, 79
    changing environments, 39
    employees involved with equipment,
        184
    maintenance management logistics,
        85–92
    matrix, 88–90
    post-maintenance era, 41
    purchase agreements, 99
    total productive maintenance, 65
    tuition reimbursement, 191
    work environment, 191
transfer, equipment
    functional setup, 171
    platform setup, 174–175
transformation and implementation,
        see also Implementation;
        Integration
    environmental studies, 200–201
    fundamentals, 5, 199–200
    goal and value changes, 202–203
    managerial preparedness, 201–202
    psychosocial changes, 203–204
    structural changes, 205–207
    technical changes, 204–205
travel, budget planning, 79
trends
    environmental, 150
    managerial, 160
    objectives, 151
    psychosocial, 158
    structural, 154
    technical, 156
triggers, 47–49
tuition reimbursement, 191

U

ultrasonic inspection, 60
universal tech personnel, 85–86, 211
UNIX operating system, 33–34, 136
unplanned interruption, 11
unscheduled down state and downtime
    fundamentals, 8, 10
    post-maintenance era, 164
    situational planning, 192
unscheduled events and defects
    preventive maintenance program
        setup, 46–47
    productive maintenance, 20
unscheduled failure reduction, 18
unsuitability, 39–40
upgrades, 37
users
    equipment user groups, 179
    information sharing groups, 51
    practicality perspective, 165
utilization
    defined, 211
    equipment performance indicators,
        115–117
    equipment performance
        measurements, 15

V

values, *see* Goals and values
variable costs, 126
vendors
    maintenance management logistics,
        95–100
    platform engineering setup,
        176–177
    technical subsystem, 155
verification run, 10
vibration analysis, 59
visual inspections

preventing failures, 60
task example, 51
Von Bertalanffy, Ludwig, 145

W

waiting, nonproductive downtime
indicators, 120
waiting work-in-progress time, 10
WAN, *see* Wide-area networking (WAN)
warranties
annual maintenance plan, 70
budget planning, 79
preventive maintenance, 19
preventive maintenance program
setup, 47
purchase agreements, 95, 97, 99
Web technology, 34, 91, *see also*
Computer-based technology

weekend preventive maintenance, 50,
83
wide-area networking (WAN), 26
wireless communication, 34–35
Wireman, Terry, 64, 105
work environment changes, 189–191
work-in-progress (WIP), 10, 140
work orders
headcount ratio, 118
management, 130
module, 132–133
per shift, labor productivity indicators,
119
real-time feedback, 121

X

XML, *see* Extensible markup language
(XML)